Shelby Marlo's New Art of Dog Training

Shelby Marlo's New Art *of* Dog Training

Balancing Love and Discipline

Shelby Marlo
Hollywood's Favorite Celebrity Trainer
with Taura S. Mizrahi

CB
CONTEMPORARY BOOKS

Library of Congress Cataloging-in-Publication Data

Marlo, Shelby.
[New art of dog training]
Shelby Marlo's new art of dog training : balancing love and discipline / Shelby Marlo with Taura S. Mizrahi.
p. cm.
ISBN 0-8092-3170-0
1. Dogs—Training. 2. Dogs—Behavior. I. Mizrahi, Taura S. II. Title.
SF431.M44 1999
636.7'0887—dc21 98-35005
CIP

Jacket photographs copyright © Petography, Inc. (Jim Dratfield/Ken Clare)
Interior design by Mary Lockwood
Interior opener photographs copyright © Petography, Inc. (Jim Dratfield/Ken Clare): 2, 14, 42, 58, 116, 164; (Jim Dratfield/Paul Caughlin): 28, 70, 96, 142, 190
Interior training photographs copyright © Petography, Inc. (Jim Dratfield/Ken Clare): 34, 36, 107, 120, 122, 124, 173–79, 184, 200
Interior photographs by Bob Dobbs: xiii
Courtesy the author: v, xix
Interior illustrations copyright © 1998 Irene Apergis: 48, 49, 54

Published by Contemporary Books
A division of NTC/Contemporary Publishing Group, Inc.
4255 West Touhy Avenue, Lincolnwood (Chicago), Illinois 60646-1975 U.S.A.

Printed in the United States of America
International Standard Book Number: 0-8092-3170-0
99 00 01 02 03 04 BG 19 18 17 16 15 14 13 12 11 10 9 8 7 6 5 4 3 2 1

To Pearl, the most loved dog ever

Contents

Acknowledgments xi

Preface: A Love Story xiii

I. Overview: For Your Information 1

1. The Big Picture 3
   What Is Training?
   Taking a History
   Top Ten Myths and Realities About Dogs
   A Few More Words of Advice
   In Short

2. Choosing the Right Breed for You 15
   A Brief Look Back
   What's in a Breed?
   Types of Breeds
   Linebreeding
   Rescue Dogs
   How Much Is That Doggy in the Window?
   Research, Research, Research
   In Short

3. Puppies, Puppies, Puppies 29
   When to Get a Puppy
   The Personal Puppy Test
   The Muzzle Grab
   Shark Mouths
   Parenting Your Puppy
   In Short

II. **Relationship: Building a Lifelong Partnership** 41

4. **What's Essential to Dogs** 43
   - Pack Animals
   - Dogs as Predators
   - Territory
   - Scent
   - Play
   - Mouthing
   - Chewing
   - Leave the Ego at the Door
   - In Short

5. **Building a Leader–Follower Relationship** 59
   - Rank
   - Rank in a Multi-Dog Household
   - Be a Benevolent Leader
   - In Short

6. **Socialization** 71
   - What Is Socialization?
   - Why Socialize Your Dog?
   - When to Socialize Your Dog
   - Entering the Social Scene
   - We Shall Overcome: Coping with Fear
   - Nature vs. Nurture
   - Problems and Misconceptions
   - A Lively Outlook
   - In Short

III. **Training: Sit, Stay, and All That Good Stuff** 95

7. **Training Theory** 97
   Evolution of Training
   Old School
   New School
   My School
   The Training Paradox
   In Short

8. **Starting Off on the Right Paw** 117
   Equipping Your Home
   Crate Training
   Dogs and Children
   Diet and Exercise
   Gentling and Handling
   In Short

9. **Housebreaking** 143
   Creatures of Habit
   Chow Time: Scheduling Food and Water
   Confinement
   Potty Times
   Potty on Command
   Will He Tell You He Needs to Go?
   Accidents
   Reverse-Housebroken Dogs
   Walks
   Submissive and Excitement Urination
   Control from Chaos
   In Short

10. **The Mechanics of Training** 165
Why Train?
What Every Dog Should Know
Have Patience
In Short

11. **Behavior "Problems"** 191
Exercise and Mental Stimulation
Separation Anxiety
Chewing
Barking
Jumping
Digging
Running Away
Other Crimes and Misdemeanors
Pica
Coprophagia
Compulsive Behaviors
In Short

Epilogue: A Final Note 217

Appendix: Quiz: Are You a Good Dog or a Bad Dog? 219

Index 223

Acknowledgments

Thanks to all who have helped shape my training philosophy, who have supported my career with dogs, and who have helped make this book possible.

To Dr. Ian Dunbar, who has been my greatest influence in my philosophy of dog training and who has changed the quality of life for dogs everywhere. To Karen Pryor and the late John Fisher for also being strong contributors to my better understanding of dogs. Laura Dern and Bridget Fonda, my first celebrity clients who really understood the training. Dr. Winters, Dr. Suehiro, Dr. Keagy, Dr. Shipp, Dr. Didden, and Dr. Fisher for their continued support. To Kali who puts up with me.

To Taura Mizrahi for helping me make sense, doing all the hard work, and putting up with all my changes. To Jacque Schultz for editing and toning me down. To Charlotte Sheedy and Regula Noetzli for believing we had a book. To Kara Leverte for allowing me to take the time to make the book right. Jim Dratfield, thanks for the great photos. To Irene Apergis for her artwork.

Special thanks to Nancy Stoddard, Spencer Cole and Henry, Kate Boleyn and Lulu, Stanley DeSantis and Aldo, Susan Harris, Paul Witt and Al, Dionne Warwick and Rio, Julie Carter, Kathy Nishimura, Joan Lachman, Rachel Murray, Lynn Lewin, Clint Rowe, Lyssa Noble, Karen Price, Carol Childs, Marcia Ziffren, Gail Zappa, Andrea King, Bob Levitan, Jim Kenny, Sandy Rendel, Tricia Tomey, Luis Vazquez, Dean Marlo, Virginia Marlo, and my mother for letting me keep Pearl.

And finally, thanks to Lotte, Ruby, Hedda, Isabella, and the chickens.

Preface: A Love Story

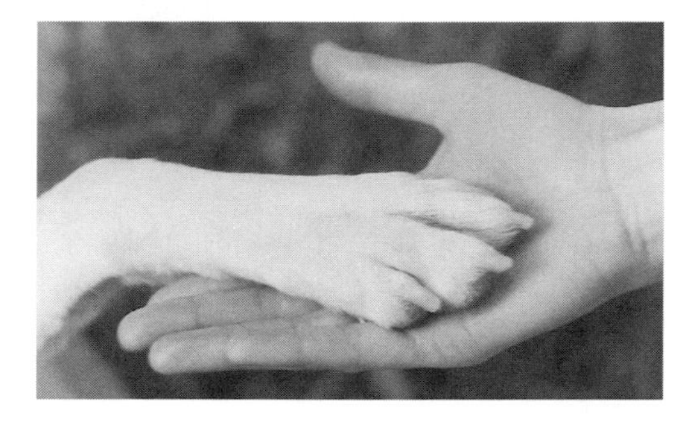

My mother was a Las Vegas showgirl. We traveled a lot, but no matter where we were I always brought home stray cats, dogs, birds, lizards, even horny toads, and I had a sixth sense for taking care of them. As much as I loved animals, moving so often made it difficult to keep a pet. My dogs would mysteriously disappear, or "run away," as my mother would often tell me with a guilty glint in her eye.

I was around nine, living in Puerto Rico with my mother, when I first witnessed true love between a person and an animal. We were on the beach when I saw this display of unconditional love between a man and his dog, Blue. Blue's owner, a surfer, swam far beyond the lagoon to catch some waves. I sat on the sand and watched Blue pine for the man to return. Nothing could console the dog, and no one could distract her. She paced back and forth on the shore, waiting for him to return from his surfing adventure. I will never forget Blue's excitement when the man did finally return.

Then came time to wash the sand from Blue's coat. Most dogs hate being hosed down, but the complete trust and communication between this man and his dog made it a simple task. He said, "Blue, turn around," and she turned, allowing him to spray her with the hose. There was no pulling or leashes, just sheer communication. Their relationship was symbiotic. I hadn't seen that before. I remember being moved by Blue's bond with her owner and knew I wanted the same. It was not until I turned fourteen that I had any control over the destiny of one of my pets. That is when I found my precious Pearl.

My mother and I had returned to Los Angeles. By this time, she had quit dancing. One day, we went to a strawberry festival in Topanga. I wandered around the wonderful, funky, little strawberry festival until I spotted a woman in the center of it all. A little, yellow puppy sat in front of her. Her littermates had all been given away. As I petted the puppy, my attention was diverted by the hippie-woman pontificating some mumbo jumbo. According to her, this was a magic puppy, not only because the father was supposedly a coyote who stole into the yard and mated with her Collie, but because the puppy was born across from an Indian reservation. There must have been a bit of truth to the magic puppy story because the puppy had to be mine. Something intense drew me toward this dog.

I knew my mother would not let me keep the puppy, but I was obsessed with having her. I sneaked her home under my shirt. It was not until my mother and I were halfway home when she heard the murmuring of a puppy. I feared my mother would turn the car around and make me return the squirming bundle tucked beneath my shirt. Maybe it was because I was older. Maybe it was because my mother and I had finally settled down. Maybe it was because Pearl truly carried some mystical, magical powers. Whatever the reason, my mother allowed me to keep the little puppy.

Training Pearl came naturally. So naturally that I was not aware I was doing it. I had always been an avid horseback rider. I took Pearl out to the ranch when I rode. As I jumped the course on my horse, Pearl followed suit. If I reprimanded the horse, so did Pearl by nipping at its legs. Whatever action I performed on the horse, Pearl mimicked. Sometimes after riding, I took Pearl to the course and had her jump it. When I went home, I set up jumps in the backyard and had Pearl run my homemade course.

I was teaching Pearl what is now called agility, a field developed from horse training. As a young teen, I wasn't thinking about training, but I *was* training Pearl. If my horse would jump, then why wouldn't my dog jump? To us, it was fun. I now realize it was much more than that. Pearl and I communicated: I spoke to her and she listened. But the respect was mutual. When Pearl needed something, I listened to her.

Around the time I finished high school, I began to seek a career. Animals were my biggest passion in life, so when I got a job as an animal health technician, I was thrilled. The best part about the job was that I could bring Pearl to work with me. For three years I worked at the animal hospital, absorbing everything I could about caring for animals. There I was able to augment my innate understanding of animals and their care with technical instruction.

Years later, Pearl started to fail. She was about fourteen. As she got older and started to lose mobility, I panicked. That's when I decided to get another dog. I knew I wanted a dog that looked like Pearl, a Collie mix. I read everything I could on dogs, both about training methods and breeds. At the same time, I took Pearl to holistic veterinarians and acupuncturists. I tried everything I could think of to help her mobility. One day while looking through a breed book, I read about smooth Collies. Then I attended a dog show where I saw one for the first time. The short-hair Collie's resemblance to Pearl was uncanny. I began contacting breeders.

By this time, Pearl had become so frail that I didn't even want to take her out of the house. I found a mobile vet to come to her. Like many older dogs, she was becoming incontinent, so I made her a bed from towels that I constantly washed and replaced. I carried her up and down the front steps so that she could relieve herself outdoors. Then out of the blue, a breeder called with a litter that had two smooth Collies.

My search for a new puppy was selfish. Even though Pearl had deteriorated past the time when a new puppy would rejuvenate her, it was something I needed to do for myself. It was the only way I could imagine coping with the loss of my first true love. When I arrived at the breeder's, one of the two puppies had sprouted hair, leaving only one smooth Collie in the bunch. Once again, I seemed destined to have only one magical choice out of a litter. I brought Lotte into her new home and immediately took her outside to her potty spot. I said, "Go potty," and she complied. We were already off to a great start.

Pearl continued to fail. I knew I couldn't hold on to her anymore. I also knew that the last place I would put her to sleep would be at the veterinarian's office. I wanted Pearl's death to be as comforting and comfortable as possible. It had to be done at home, so I called the mobile vet.

Euthanasia was one of my jobs when I worked at the veterinary clinic. For me, the hardest part about putting animals down is the last look they give you before they pass on. I wanted Pearl to go peacefully in her sleep, so I served Pearl her favorite meal laced with Valium that the vet had prescribed. By the time the mobile vet arrived, Pearl was asleep. When the veterinarian gave her the injection, Pearl did not move—not even a flinch. Pearl traveled from sleep to complete relaxation, and then she was gone. After Pearl's funeral, I came home to Lotte. Had I returned to an empty house, I am afraid I would have fallen apart, but Lotte helped me hang on to the routine of my life.

Once it was just little Lotte and me, I got to thinking that maybe I should formally train her. I had seen a sign in the park for classes and called the number. When I was younger, the word *train* never entered my mind, let alone a class. Because I was older, I had lost faith in my instincts. Ironically, it took that first class to renew my faith in my natural abilities.

The first day of class, I stood attentively with Lotte by my side and listened to the dog trainer tell the class about choke chains. Choke chain? What was that? Then he spoke about shock collars for aggressive dogs. The things he said were entirely foreign to me. I had been a natural trainer my entire life, and now this man wanted me to jerk this little puppy I loved so much. No way. Although I had paid quite a bit of money for this man's expertise, I ignored much of what he said.

I signed up for more classes. I was enrolled in three different obedience classes at the same time. However, they all used the old-style, coercive training methods. Lotte was learning. How could she not? Training was all I did. Her attitude was poor, though, and I resisted a lot of what I was told. It didn't feel right. Lotte, once so happy to learn, was slowly changing. There was nothing wonderful about the training sessions. She was becoming a completely different dog in class compared with the spirited dog I played with when we were alone in the park. But I persisted, and we continued to train.

All that time, I had only encountered compulsion methods of training. It was not until I became a member of a local dog club that I learned about motivational training and the use of food. Initially, I thought using food was bribery and resisted the idea. I wanted my dog to work for me, not for food. This was also the first time I learned of Dr. Ian Dunbar and his training methods with food. I read everything he had written, and the more I learned, the more the idea of using food as a reward began to permeate my brain.

While I continued to attend club events, I sifted out people who could help me learn about this other type of training. I was also getting ready to try my hand at competition obedience trials and was referred to a trainer who, like a coach, helped people prepare their dogs for competition. I imagined Lotte and I would parade around in front of her while she critiqued us, and that then we would be happily sent home ready to compete. My world exploded when she gave me the news.

Lotte performed, but our coach quickly informed me that I had no attention from my dog. She said I could spend the next year retraining everything motivationally, or I could spend three months doing a patch-up job. At that point, I had already worked so long with Lotte and was so anxious to start competing that I could not face the idea of another year of training. Over the next three months, Lotte and I met with our coach twice a week.

Lotte was superb. We trekked uphill, continually running into coercive-style trainers and books. Then, suddenly, I saw the valley. We were exposed to an entirely new world of training with food. The day I brought food into training was the day I brought my smooth Collie back from the dead.

I learned the power of positive reinforcement. I learned about conditioned reinforcement; for example, saying "good" before giving a food treat so that later "good" alone has a very powerful meaning. I learned to vary my use of food to create maximum excitement. I learned to shape behaviors, giving the food treat on the very best effort. I learned the idea of the jackpot: rewarding a breakthrough with lots of treats so that the effort sticks in the dog's mind. I learned to stop on the best effort, resisting the strong temptation to try one more time. I learned the importance of timing, immediately rewarding correct behavior and reprimanding undesired behavior.

For the first time, learning together was exciting and lively. Lotte's understanding was quick, strong, and sharp. We were a team. By the time we entered the ring at the Beverly Hills Kennel Club dog show, Lotte was a completely different dog. We won the Novice-A division, but I had no idea exactly how well we did until the end of the day when we were awarded High-in-Trial with an outstanding 199 out of 200 points. Lotte and I had placed higher than anyone else at the competition that day, including our mentor.

As teammates, Lotte and I proceeded to win every competition we entered—including the Collie National where we scored a 198 and another High-in-Trial. Lassie presented Lotte with her award that day. From that point on, my career took on a life of its own. I started working with a veterinarian who referred me to his clients. Since then, I have amassed a large clientele, appeared on many television shows, including *The Oprah Winfrey Show*, and have been profiled in a variety of magazines.

During my time in the dog obedience world, I have learned many things. Most important of all, I have learned to be open to different methods. There is no single answer in training; it is a learning experience for both you and your dog.

I have come a long way from that beach in Puerto Rico where I first watched in awe true love between a person and an animal. I don't have to search for that love anymore. I had it with Pearl. I have it with Lotte. My goal now is to help others find that pure love with their dogs.

I hope I can help you find your Blue.

Me as a kid on the beach in Puerto Rico with Rhonda, my foundling

Shelby Marlo's
New Art *of*
Dog Training

Part I
Overview

For Your Information

1

The Big Picture

There is an incredible movement happening in the dog world. A kindness revolution. The old-style training is finally being put to rest, and the new is struggling to understand itself. The pendulum has swung from the extreme of using only negative reinforcement to the extreme of using only positive reinforcement. I have studied both training styles, and it is clear that the new school is far superior. However, when appropriate and applied properly, a correction can be an extremely effective tool. Used together, the two styles bring a balance of love and discipline to training. It is time to steady the pendulum.

What Is Training?

Training is more than just sit and stay. It's also learning about dog behavior and the canine-human relationship. Once the owner has been initiated into this way of thinking, we are ready to train the dog. First, you must get a sense of the dog's temperament. Dogs, like people, have their own personalities and need to be treated individually. A brash, tough dog must be trained much differently than a shy, timid dog.

The universal common denominator with dogs is almost always the almighty treat: the quintessential social lubricant! We may expect our dogs to be on a higher plane, but face it, your dog responds to treats! Don't take it personally. Be happy you have discovered a motivational tool. Once in a while, I find dogs that work better for a toy. They tend to be terriers and herders. But even those dogs will sometimes break down and work best for the right treat.

Dogs are also emotional beings with which we have some wonderful spiritual connections, like allegiance to family. Because dogs are pack animals and social beings, *you* are the center of your dog's universe. People sometimes confuse their dog's dependence with the desire to please. The reality for many dogs is once they are secure with their surroundings, the desire to please goes out the window.

The "Love, Praise, and Affection" Fallacy

Beware of trainers who use so-called "love, praise, and affection" training. They employ choke chains instead of food rewards. This is how it works: You tell the dog to sit. The dog has no idea what sit is so you jerk the choke chain while pushing the dog's butt down. Soon the dog learns that if he doesn't sit, he will be reprimanded. Upon compliance with the correction, the dog is patted on the head and told "good dog." These trainers espouse that the dog is working for a pat on the head, when it really works to avoid punishment. Do you see the irony?

These trainers feed into the myth of the desire to please under the guise of love, praise, and affection, a.k.a. a pat on the head. The desire to please DOES NOT EXIST. Instead, the dog works to avoid a severe negative punishment. He is not working to please the owner; he is working to avoid a harsh leash correction. He does not do it just because he loves you or because he's a good dog; he does it because he is frightened of what might happen if he does not do it.

Relationship

When I call for my dog Lotte, she comes to me. At this point she does it without the lure of a treat. In fact, she knows that the probability is high that she won't get a treat. So why does she come to me? True, she's been conditioned so many times that somewhere in the back of her mind, she's thinking there is a possibility she will get a treat for coming to me. At some point, relationship takes over, and she comes to me no matter what. We have a strong leader-follower relationship.

Lotte does not come to me to please me or satiate some need of mine. I accept that. Dogs do not have a complex, abstract way of thinking. If you distill a dog's behavior down to its essence, you will find the root of Darwinism: the survival of the fittest. A dog obeys his leader to get something for himself. He either receives pleasure from gaining a reward or avoids pain in the form of a reprimand.

A strong leader praises as well as sets boundaries. Don't feel you are breaking the dog's spirit by giving him rules. When a dog has a strong leader, he feels safe and secure that his pack will go on. Therefore, dogs are naturally attracted to a benevolent leader. A leader to a dog represents strength, security, and survival. Lotte comes to me because we have a solid leader-follower relationship.

Taking a History

When I first meet new clients, they always lead me straight to their dog. I lead them to the kitchen table, and I take a history of the dog. Through this information, I know how best to serve my clients and their dogs. After obtaining the dog's history I begin to train the owners. They must learn what they've been doing right, what they've been doing wrong, and how best to change it. Training is about more than just the dog; it's about the owner, too.

One of the first questions I ask when I see new clients is, where does the dog sleep at night? This question gives me great insight as to how these people relate to and feel about their dogs. The answer can run the gamut from "he sleeps in bed with me," "he sleeps in a crate in the bedroom," "he sleeps in a crate in the laundry room," "he sleeps wherever he wants to," or "he sleeps outside."

Questions and answers such as these shed light on how people view and deal with their dogs. I need to know how my clients relate to their dogs so I can get a better sense of where the dog's behavior problem stems from. Are you paper training your dog? If so, STOP! You are teaching the dog to eliminate in the house. Are you punishing your dog *after* he has engaged in a "bad" behavior rather than *during* the behavior? If so, STOP! You are teaching your dog to fear you and to do the behavior when you're not around. Your dog will chew on the couch when you're in the other room, and he will urinate on the rug when you're not home.

With a little forethought, it's easy to give dogs acceptable alternatives to their normal canine behaviors. We teach our dogs what is acceptable and unacceptable behavior within our household. We take into account that we are responsible for showing our dogs what is correct behavior. Why wouldn't the dog chew the couch? What else is it there for? It is our responsibility to teach the dog that it is not okay to chew the couch, but it is okay to chew a rawhide chew toy. We offer appropriate, alternate outlets for normal dog behaviors. Eventually, eating the couch is no longer an issue.

Top Ten Myths and Realities About Dogs

The following are some popular myths about dogs:

Myth 1. Don't start training a dog until he is six months old.

Common belief was that a dog could not be trained until he was at least six months old. In actuality, the delay was not because the dog was not trainable but because of the methodology used. Training used to be a very negative and often harsh process and thus was not advised for the young—and should not have been for dogs of any age for that matter. Training methods were so harsh that a dog had to be of a certain age to withstand that level of abuse. Waiting to train had nothing to do with the dog's learning ability. We now know that adolescence is the least desirable time to begin training. At six months a puppy has already undergone a lot of negative learning and is now at the onset of puberty. If a puppy is old enough to leave its mother and littermates, it is old enough to be trained with kind and gentle methods.

A puppy is as capable of learning as an adult dog. At forty-nine days, a puppy's brain waves are the same as those of an adult dog. Puppies are little sponges soaking up every bit of information they come in contact with. If you do not show them what you want them to do, they're going to figure out their own way. They learn that it's fun to chew the couch. They learn that urinating on the carpet brings relief. They learn whether you are involved or not, so you might as well be involved and teach correct alternatives to their natural, little-puppy inclinations. Still, some vets, breeders, and wise old dog people are remiss in their failure to recognize this.

Myth 2. Keep the puppy behind closed doors until he's had all his shots to keep him safe from diseases like parvovirus.

Keeping the dog behind closed doors is not the panacea you think it to be. Unless this dog is living in a bubble, he will never be safe from exposure to disease. You can step in animal waste and inadvertently track it into your home on the soles of your shoes. In fact, isolating your young puppy from the world could foster more harm than good. If you don't socialize your puppy prior to three months of age, you run the very real risk of creating a fearful, antisocial dog.

By taking calculated risks like going to a puppy class where all participants are vaccinated and healthy; visiting with friends who have healthy dogs; taking the puppy for drives in the car on errands; and having puppy parties to which you invite acquaintances of all colors, sizes, and ages, you will create a well-rounded, confident adult dog that is worldly in his views of the human realm.

Myth 3. The puppy chews on you because he is teething.

Yes, the puppy may be teething. More likely he is treating you like another puppy, and he's exploring his world and boundaries with you. You must teach your puppy not to place his teeth on you. Do not indulge him because you think it is something he needs to do. You are not his chew toy. Some mouthing is necessary to help teach bite inhibition. Otherwise you could end up with a dog that has a hard bite—and a hefty lawsuit.

Myth 4. Don't neuter or spay your dog because it will get fat, lose its desire for life, and suffer severe personality changes.

Dogs get fat for the same reason people get fat: too much food and not enough exercise. The specific changes in a male are simply due to the fact that he no longer has a strong desire to roam, fight, mount, and otherwise exhibit undesirable behaviors. If you feel these behaviors are not a problem with your dog, then neuter him to prevent prostate and testicular cancer. Females that have been spayed before their first estrous cycle will not develop uterine cancer, breast cancer, or other female ailments. Spayed or neutered dogs have been shown to live longer, healthier lives.

Obviously, spaying and neutering your dog helps put an end to the tragic overpopulation of dogs. More than six million animals are euthanized in pounds and shelters every year. This figure does not include the countless number of dogs that are killed on the street or starve to death. We spend over one billion dollars every year to euthanize dogs, yet we blithely call them "man's best friend." Even if you decide to breed your dog and find the puppies homes, you are still not exempt. Those homes you found for your puppies were homes that would have probably adopted another dog already in need of a home. You merely add to the surplus, the grief, and the heartache by breeding your dog. Leave breeding to the reputable hobby breeders.

Myth 5. Dogs are little people in furry coats.

Dogs are very different from people. They might as well be aliens from another planet. While dogs do share some of our similarities—being social animals, having the ability to care for their young, and developing strong bonds—they do not share our ability for abstract thought. Dogs do not think about the past or future or feel guilt or vengeance. Dogs do not share our moral codes. This does not make them bad creatures, merely different. Problems occur when we wrongly impose our views on them. We expect too much of our dogs. We must accept the fact that dogs are not little people in furry coats. We need to understand that what we value has no bearing on what a dog values.

Myth 6. Dogs have a desire to please.

Dogs, like every other organism, are driven by survival. The fact that your dog does not have a desire to please you does not mean you cannot have an intense, loving relationship with him. If your dog does not come when you call him, it's not because he doesn't love you. It's because you have not properly reinforced coming or because a competing motivation is more powerful. Strengthening a command entails finding a powerful motivator such as a treat or a toy.

Myth 7. Crates are like jails—ostracizing, mean, and cruel.

Dogs are den animals by nature. They tend to seek small, confined spaces like under a bed or table or in a cubbyhole. To a dog a crate is safety, security, and a place of his own. Crates are incredibly useful tools for housebreaking as well as preventing destructive behaviors like chewing. When placed in a high-traffic area, the dog still feels he's part of the family. At some point in your dog's life, he is going to be placed in a crate whether at the vet's, at the groomer's, or while traveling. It's always a good idea to acclimate the dog from the start to avoid future trauma.

Myth 8. When the dog does something wrong, just say "no."

Certainly a dog can be trained using the word *no*. However, it is more educational, fair, and easy to use an instructive reprimand. This means using a word that not only tells a dog to stop what he's doing but also instructs him in what he should do. For example, if the dog is caught eliminating in the house, you say "outside," teaching him that potty is a location issue and therefore he should eliminate outside. If the dog is jumping up on a person, you say "off," telling him he should get off the person. If the dog is barking, you say "quiet," and he should become quiet. Instructive reprimands are specific to the current behavior or action. Your dog may stop when he hears "no," but he still has to pee. An instructive reprimand gives him a correct direction.

Myth 9. You cannot be indulgent with your dog and have him be well behaved.

You can be as indulgent as you want with your dog as long as boundaries and training are in place and he views you as a leader. My dogs are allowed in my bed and I feed them from the table. I do not have problems with my dogs because I am a strong leader. If I want my dogs to get off my bed, I tell them to get off and they do. If I don't want them hanging around my dining room table while I'm eating, I simply tell them to go lie down and they do. These are privileges that I bestow upon my dogs, and I take great joy in doing so without losing their respect.

Myth 10. Dogs know when they've been bad.

Dogs don't think in abstract terms, and guilt is an abstraction. If your dog's ears are back, his tail is tucked, and he has an overall low body posture, he may look guilty. We misinterpret the dog's slinking as a sign of guilt but that isn't what he's feeling. People falsely believe the dog thinks "if only I hadn't chewed the couch."

Your dog assumes a submissive body posture as a direct reaction to your signs of anger. When you come home and punish your dog for chewing the couch, he does not associate his act with his punishment. He associates his punishment with your homecoming and learns to fear your arrivals. This can exacerbate the situation because he becomes anxious about you coming home and chews the couch even more.

A Few More Words of Advice

Choose your breed or breed mix wisely. Breeds were developed to fulfill different functions, and they may not always be appropriate for our lifestyles. If you bring a dog into your home that does not fit your lifestyle, then you may be facing a lifelong struggle with your dog.

Do not take what I say lightly, for many of the dogs that reside in shelters or rescue organizations are dogs that were matched with the wrong home. They are not bad dogs; their specific breed traits just did not meld easily into the household and were viewed as behavioral problems that took too much effort to combat. For instance, a very active dog will not fit into a sedentary household. Consequently, these dogs are given up for adoption . . . which leads us to "rescue dogs."

A rescue dog is any dog that has lost his first owner after initial placement. Because these dogs have lost the security of their pack or family, and because to a dog *pack* always means survival, they have a grave need to bond with another person or pack. They are at a higher risk to overbond with their new owner and give in to displays of separation anxiety. The dog will not leave the new owner's side. Some new owners are flattered by this response, but it is not healthy for the dog. This is not a unique situation. Setting boundaries will help the dog form a sense of security and permanence in his new household. Don't feed into his insecurity and think you can love him back to health. You will cause more harm than good.

In Short

There is a revolution going on in the world of dog training. By using positive reinforcement and appropriate corrections, you get not only a dog that *wants* to follow a command but one that understands that you are the leader and he is the follower.

In order to get compliance and build your dog's desire to want to work with you, discover what motivates him. Food is usually the strongest attention-getter and most powerful reward. Beware of people who criticize you for training with food. They often train only in a compulsive manner. Strict compulsion training may create a dog that works solely out of fear.

Dog training and problem solving are more than teaching a handful of commands. Taking a history of the routines and methods the pet owner employs and how the dog responds to them reveals the human-animal relationship, and a training protocol incorporating a strong leader-follower relationship can be developed.

There are certain old wives' tales and myths that stand in the way of dog owners' success. Look for the truth behind these myths to ease your way.

The following are some old wives' tales dispelled:

1. The optimal time for a puppy to begin training is at forty-nine days.
2. Puppies need to be socialized with prudence even before they are fully vaccinated.
3. Puppy mouthing is more than teething.
4. Spaying and neutering does not diminish your dog's desire for life.
5. Dogs are different from humans and have a separate set of codes and values.

6. Dogs do not have an emotional desire to please.
7. Crates can give a dog a sense of safety and security and facilitate housebreaking.
8. When the dog does something wrong, give him an instructive reprimand such as "quiet," "off," or "outside."
9. If boundaries and training are in place and you have a strong leader-follower relationship, you can be indulgent with your dog and he'll still be well behaved.
10. Dogs don't feel guilt for inappropriate behavior.

Lotte and Ruby

2

Choosing the Right Breed for You

When looking for a dog like Pearl, my aging Collie mix, I researched and researched until I found what I wanted. One thing I had learned through my research was to ask the breeder about any particulars regarding temperament in the dog's line. I was perplexed when the breeder told me of Lotte's hesitation to traverse linoleum floors. Obviously, a trait like this would have proven problematic had my home been floored with linoleum or hardwood. It is crucial to consider these types of issues when looking for a pet that fits not only your lifestyle but your home as well.

Man has bred dogs to do everything from finding and killing rats to pulling sleds through the snow. They herd sheep, guard flocks, point hunters to game, and sniff out criminals. Different breeds serve different functions. They also have different physical features to assist in their individual functions. Darwin called it evolution. Since few of us actually live with working dogs, some breeders have turned to breeding for looks rather than working ability. This has been problematic in some breeds, resulting in heads too large for birth canals, coats too thick for warm temperatures, and other features that run contrary to survival.

A Brief Look Back

Even though you may think your new Schnauzer, Ralph, is just like you or me, he's not. No matter how much we treat our dogs like people, they were, are, and will always be descendants of wolves. It's taken about twelve thousand years for the dog to diversify into roughly four hundred breeds. It's also a fair bet that domestication took place simultaneously in different parts of the world.

Man's best friend did not always hold this title. At one time, humans and dogs competed for food. When people were hunters and gatherers, they traveled in packs like dogs. They stalked much of the same prey as dogs. Imagine, your ancestors and Ralph's ancestors probably squabbled over the same rabbit. Not only did humans hunt in a manner similar to that of canines, they also had similar social systems: both established a central home base with a system of defense, both lived within a family structure, and both took care of their young.

What is probably the most important similarity between the two packs, human and canine, is each established a hierarchy, starting with a leader and working down based on a division of responsibility. To canines, the alpha, or "top dog," was like a king. He had the best sleeping location, ate first, was the only one to procreate, and had the choice portion of the food. No one messed with the top dog, and each animal related to the others depending on where he, himself, landed in the hierarchy. Even today, dogs see most everything in terms of rank relevant to themselves.

Eventually, humans became more sophisticated, and so did dogs. As the two packs evolved and learned to live together, humans became top dog. By the Middle Ages, they were breeding dogs of like characteristics, traits, and behaviors to create dogs that specialized in guarding property, herding and protecting flocks, and even assisting people in hunting. Our competitor became our companion. As dogs learned to depend on and work for us, their place in the pack was clear. Humans had become the leader of the pack.

As humans began to travel and explore the world, they brought their dogs with them. They created new breeds with new jobs. Dogs were no longer bred solely for function, but for appearance as well. We wanted to show off our dogs, so we put them on display. Cities continued to grow and people were no longer living in extended families. Dogs took on a new role as "man's best friend."

Recently, dogs have taken on an even more intimate role; they have become family members. Unfortunately, many people confuse loving with spoiling. In our modern lives, we have taken away the clear distinctions of a pack hierarchy. Herein lie our problems with training our dogs.

What's in a Breed?

Technically, a breed is a group of dogs that share a fixed conformation, like ear or muzzle shapes, which allows them to do a specific job efficiently. Each dog within a breed tends to share behaviors, too. Retrievers carry things in their mouths to their owners. Hence, their name. Collies are bred to herd. Instinctively, they fix their eye on a moving target, position themselves behind it, and bark at it. Tibetan Mastiffs were trained historically as guards of both the flock and the home. Expected to sit alone hours on end protecting the flock, these dogs are by nature independent, aloof, and aggressive.

Ruby

When I became a home owner, I wanted a dog that would make intruders think twice about robbing me. A recent movie, ironically called *Man's Best Friend,* about a genetically engineered dog that flips out and kills everyone around him, featured a Tibetan Mastiff. No burglar would dare invade a house armed with such a security system. The dogs used for the film had black and tan coats, long hair, and massive heads to go along with massive bodies. They definitely looked the part. I wanted one of my own. And so I got Ruby.

When choosing a breed, you have to decide what is important to you and then look at what the dog is bred to do. When I picked Ruby, I wanted a big guard dog that looked menacing to strangers. She fits the description. I like a dog that is obedient. Having said that, however, I have an inherently independent dog that wants to attack 50 percent of the dogs she encounters. She is trained perfectly off-leash. But because of the nature of her breed, I cannot take Ruby for off-leash walks in the presence of unfamiliar dogs. So, I accept my limitations with her, including rearranging my schedule to hike at dawn, when we are less likely to encounter new dogs.

A Breed for You

Frank and Barbara Sinatra got a Weimaraner. This young, active dog had no business being in their home. Even though the dog was a sweetheart—affectionate, lovely, easily trained—it was a complete mismatch with the Sinatra household. She followed commands and had a very responsive off-leash recall, but all the training in the world could not drain the energy from Shadow, the Sinatra Weimaraner. Shadow was a whirling dervish that ran through the house and up and down the stairs, taking out anyone in her way. Barbara finally realized the potential danger this posed. She understood that her home was the wrong place for this sporting dog. Eventually she gave Shadow to a close friend whose lifestyle better matched the dog's.

Barbara Sinatra was lucky to find an appropriate home for Shadow. Many people choose a breed without really learning about the exercise needs and behavioral challenges inherent to that breed. Probably the most important question to ask yourself when choosing a dog is, *what was this dog bred to do?* Then compare the answers with your lifestyle and level of dog-owning experience.

Types of Breeds

For the most part, a breed will fall into one of two categories: those bred to work with people, and those bred to work independently. Generally speaking, the dogs that are easiest to train and best take instruction are the herding dogs and the sporting, or gun, dogs. The master instructs the dog to retrieve, to herd, to return. Other dogs, those bred to work independently, are the flock guard dogs, the terriers, and the hounds. These dogs tend to be self-directed. They don't rely on humans for instruction. This trait is instinctual.

Please keep this in mind when choosing a dog. Clients complain to me all the time that their dog is not responding the way they had hoped. For example, a family purchases a Jack Russell Terrier to be their companion, but the dog digs up their garden. Terriers were bred to hunt and kill vermin. They go to ground to flush out their quarry. The dog is merely following his strong, natural instinct to dig. Knowing what a dog was originally bred to do must be a factor in your decision. This is one benefit of bringing home a purebred, since physical traits and behavioral characteristics are usually more predictable.

There are several good books on dog breeds. I suggest *The Right Dog for You: Choosing a Breed That Matches Your Personality, Family and Lifestyle* by Daniel F. Tortora, Ph.D. (Fireside, 1980) and *Your Purebred Puppy: A Buyer's Guide* by Michael Lowell (Henry Holt, 1990).

Rare Breeds

Rare breeds may be novel and have unique appearances, but beware. Not only are they hard to find, but information and resources may be scarce. Vets, groomers, and trainers may have limited experience with them and may not recognize normal and abnormal breed traits. Also, because there are fewer of a particular breed, the gene pool may be extremely small, increasing the risk for genetic problems.

Crossbreeds

Crossbreeding is the mating of two different purebreds. Many people may think, "Great! The best of both worlds." But be careful. It can also be the worst of both worlds. I once worked with a cross between a Rottweiler and a Golden Retriever. Not only did the poor thing have hip dysplasia from both parents, but mentally, he was a basket case. Part retriever, he wanted to play and retrieve all the time, but he inherited aggressiveness and possessiveness from his Rottweiler parent. The dog was so extremely conflicted between playing and fighting that he was a wreck. When buying a dog, there are no guarantees.

Mongrels

The mongrel is more commonly known as a mutt, mixed breed, or Heinz 57 dog. It is a mixture of several breeds brought together. Generally, one of the breeds predominates the look of the dog. These dogs tend to be bright and personable when well trained and properly socialized.

Natural Dogs

Man has created diverse breeds from the giant Great Dane all the way down to the teacup Chihuahua. However, the breeding of natural dogs is not controlled or manipulated by man. Natural dogs generally develop a very specific look. The pariah dogs seen around the world are all about thirty-five to forty pounds with short hair, prick ears, a medium muzzle, and a long tail which sometimes curls up. Natural dogs, such as the feral dogs roaming the streets in many underdeveloped countries, are closer in look to a coyote or wolf. If all the dog breeds were collected into one space and allowed to breed at will, they would eventually get back to their natural look. Try to find a dog with several of these natural aspects. It's my bet that it will probably be more physically sound.

Linebreeding

Dogs, like anything in nature, evolved to fit their environments. Each breed has certain behavioral and physical characteristics which meet the demands of its intended tasks. We have manipulated their functional appearances into a "look." Because a breeder cannot be absolutely certain what the outcome of mating two dogs will be, many linebreed their dogs in order to increase their chances. Linebreeding is a form of inbreeding directed toward keeping the offspring closely related to a superior ancestor. A more extreme-looking and beautiful dog is more likely to catch the judge's eye and bring home a trophy. Unfortunately, beauty comes at a cost. Many of the resulting dogs suffer from physical and psychological problems.

Case study: Walter, the Bulldog. Cute as he was, Walter was a walking wreck. He couldn't breathe. He couldn't do anything. Like almost all Bulldog puppies, Walter was born via cesarean section because his head and shoulders had been made so big and broad by man that they were too large to fit through the mother's birth canal. Walter is a living example of the fact that dog breeds are man-made. The flattened face, man-made. Flop ears, man-made. Big, centered eyes, man-made. The Bulldog is about as far removed as you can possibly get from the natural dog.

Rescue Dogs

A rescue dog is any dog, purebred, crossbred, or mongrel, that has lost his first home. Rescue dogs may be problematic because when a dog has lost his first family, he tends to suffer like a child taken from his home. This does not mean your new seven-week-old puppy is considered a rescue dog. At that young age, he is still considered an infant, and he accepts new people as extensions of his family pack.

If you feel strongly about not perpetuating the breeding of dogs or do not want to deal with training a puppy, then you would most likely decide on a rescue dog. There are definitely far too many dogs without homes. However, a rescue dog comes with certain risks. The main risk is not knowing how the dog was raised in his previous home.

Rescuing a dog can work out wonderfully, but there are times when it does not. Be prepared for anything. Try to learn as much about the dog's history as you can before bringing him home. Was he turned in by an owner, or was he a stray? Try to hedge your bet by checking into what this particular dog has been through. It's also wise to bring an experienced dog trainer to help you choose your rescue dog.

Sometimes you get lucky. One of my most beloved dogs is Huggie, a female Shepherd-Spaniel mix rescued from a Los Angeles pound by singer George Michael. Huggie was approximately three and a half to four months old when George brought her into his home. Huggie loves life and is a natural at everything. She was easily trained and quick to overcome any fears she had already developed, like a fear of riding in the car. She acclimated to sleeping in the crate and quickly learned how to play with other dogs. Huggie is an all-around wonder dog. We knew nothing of Huggie's background other than that she was found as a stray on the street, and she turned out to be brilliant.

How Much Is That Doggy in the Window?

Not many of us can pass by the pet store in a mall and resist cooing at the puppies in the window, but you may want to think twice before purchasing a pet-store puppy. While many pet-store puppies work well in the home, you perpetuate the barbarity of puppy mills when you buy a dog from a pet store.

Dogs are naturally very clean animals. They do not soil where they sleep or eat. However, most pet-store and puppy-mill puppies are forced to function against their natural instincts. They are kept in small confined areas that serve as living, dining, and toilet quarters. A puppy raised in such conditions usually becomes what I call a "dirty puppy."

A dirty puppy is a dog that has no inhibition about urinating and defecating where he sleeps. This is problematic because you now have a dog with built-in housebreaking problems, which is not exactly the free-gift-with-purchase you were hoping for. Pet owners will have a hard time trying to housebreak puppies that have been forced to ignore their natural instincts. To avoid these problems, check out the puppy's living conditions before you bring him home. This is a good rule of thumb whether you are shopping at the pet store, dealing with a breeder, or seeking to save a life at a rescue society or shelter.

Research, Research, Research

Although the urge may be strong to pick up the first cute little puppy you see and bring him right home, you must RESIST! It is imperative that you find a dog that meshes with your lifestyle. There is so much to consider before choosing a dog: where you live, how much time you have to devote to the dog, whether or not you have children, what purpose the dog will serve in your household. The list goes on.

Many breeds have particular health problems to consider as well. Once you select several breeds, research their health problems. There are several ways to find out which breeds suffer what problems:

- Read an excellent reference book called *Medical and Genetic Aspects of Purebred Dogs*, edited by Ross D. Clarke, D.V.M., and Joan R. Stainer (Forum, 1994).
- Talk to reputable trainers, groomers, and veterinarians who will tell you the truth about a breed, not just what you want to hear.
- Go to a dog park and talk to the owners of that breed. Don't be shy to ask about their general experiences with the dog or the dog's temperament and health.

Where to Find a Breeder

When you have decided on a breed and are finally ready to find the dog for you, contact the American Kennel Club, or AKC (5580 Center View Drive, Raleigh, NC 27606), and ask for the name and number of the secretary of a breed club in your area. The secretary will give you a list of local breeders. I found Lotte by contacting the Smooth Collie Club secretary in my area. Do not be afraid to interview the breeders. Find out to whom they have sold dogs, then contact those people and find out everything you can about the dogs they got from the breeders. Pick several breeders who sound the most reliable, and go see their premises.

Breed Rescue Organizations

If you don't have a lot of money to spend or want to save a dog, and you really have your heart set on a particular breed, you might want to contact the rescue organization for that breed. Every breed has a rescue organization. Even some of the more esoteric breeds have their own rescue organizations. They are usually run by people who are in love with and particularly knowledgeable about the breed. Each November the AKC *Gazette* lists names and phone numbers for the rescue coordinators of each national breed club. These people, in turn, can direct you to local rescue organizations. Many animal shelters can do likewise.

This is a fantastic way to help a dog and get the breed of your choice without mortgaging the house. The one drawback: these are rescue dogs, and you never know what kind of excess baggage they could potentially bring into your home. The biggest plus: they could be the greatest miracle of your life.

The Third Degree

Your dog search does not end with simply choosing a breeder. Once you actually meet the breeder and before you choose a dog or puppy, there are specific questions you should ask the breeder:

- **Was the dog linebred or a complete outcross?**
  You want to find out if the dog's father is also his grandfather (linebred), or if the parents come from two completely different ancestral lines (outcross). Keep in mind that an outcross helps, but does not guarantee against the expression of genetic defects.

- **Are the dog's parents free from the breed's genetic problems?**
  Ask for documentation which states that the parents and, if possible, the puppy, are clear of any genetic health problems. When applicable to the breed, ask for the Orthopedic Foundation Association (OFA) certification number to help ensure against hip dysplasia and the Canine Eye Registry Foundation (CERF) number, updated annually, to ensure healthy eyes.

- **What age were the parents when bred?**
  You don't want a dog bred on the first heat because the mother is physically immature. It is best to wait until the dog is at least two years old to breed it. At that point, the dog is more physically and mentally prepared for breeding. Also, a dog cannot be OFA certified until the age of two.

- **What titles does the dog hold?**
  You might want to find out if a Golden Retriever has field awards for chasing ducks or birds, or a Miss America crown for a pretty coat. A field stock dog that works for a living may be too high-energy for most urban lifestyles. Look for breeders who have successfully tested their dogs for the Canine Good Citizen (CGC) title, who involve their dogs with pet-assisted therapy, or who work their dogs in agility. These activities give you a better indication of the dog's temperament and trainability.

- **What efforts have been made to socialize the dog?**
  Watch the dog for a while, and notice how he interacts with people and with his littermates. If a dog has not properly interacted with humans or other dogs in the early, formative weeks of his life, he could develop behavior problems. He may also suffer from kennelosis, a symptom of fear found in a dog that is only comfortable in his kennel environment.

- **How was the dog nurtured in the first few weeks of his life?**
  If possible, ask to see the parents of the puppy. Not only will this give you a hint as to what the dog will look like when full grown, but it's a good chance to find out how social or aggressive the parents are.

- **Why did the breeder breed the dog? For money? For fun? To further the breed? Is the breeder keeping any of the puppies?**
  You want to know what the breeder's interest was in breeding the dog. If a breeder wants a puppy from the litter you are looking at, that says a lot about the breeder's personal confidence in that litter. However, realize that a very good, well-established breeder may have a long waiting list and also may not have the capacity to keep a puppy from every litter. The best breeder is one who takes responsibility for the puppies she produces: socializing them, keeping them in a clean environment, having them examined by a veterinarian and inoculated, and stipulating in her contract that the dog can be returned at any point during his lifetime if things don't work out.

Stand Your Ground

If the breeder refuses to answer any of your questions or turns down any of your requests, hightail it out of there. These are all valid questions which a trustworthy breeder shouldn't mind answering. In fact, a good breeder will grill you nearly as much as you grill him. Don't feel like a nuisance, and be aware that the more you know about the breed, the more confident you will feel when interacting with the breeder. By doing your research, you will seem more credible, so arm yourself with questions and stay strong. Remember that all dogs and puppies are cute, and it's more important to find the right one for you.

In Short

Over the last twelve thousand years or so more than four hundred dog breeds have evolved from a common wolflike ancestor. Through the centuries, the role dogs play in our lives has evolved from competitor to helpmate to companion and even family member. As humans selected and bred dogs for certain physical characteristics that enhanced working abilities, breed types became fixed and distinguishable from one another.

When choosing a dog, it is important to recognize the limitations of your lifestyle and then select the breed that is best suited to live within those limitations. It is essential to understand what a breed was originally bred to do and whether the dog you are considering holds any titles for that trait.

Almost all purebred dogs can be divided into two categories: dogs bred to work with humans and those bred to work independently. Sporting or gun dogs and herding dogs fall into the first category, while flock guard dogs, terriers, and hounds fall into the second.

If you are thinking of obtaining a rare breed, consider the following: professionals may know little about how to treat your dog, the dogs may be hard to obtain, and dogs from small gene pools may have greater health risks due to inbreeding.

For some, a crossbreed, which has two purebred parents of differing breeds, or a mongrel may fit their needs. While mixed breeds may not be as predictable as purebreds, their unique look and reasonable purchase price may suit you. Most crossbreeds and mixed breeds are as bright and personable as any purebred—and some are even more so, if well socialized and obedience trained.

When considering a rescue dog, think about whether the merits outweigh the risks in your situation. Since there is usually very little history available, you may feel more comfortable with a professional trainer at your side who is able to do a temperament evaluation for you prior to adoption.

Before making any selection, do your homework. Whether you are headed to a breeder, rescue group, or shelter, come armed with a list of questions. And be prepared to answer as many questions as you ask, for a good fit is best for all concerned.

3

Puppies, Puppies, Puppies

At forty-nine days, a puppy's brain is fully formed and receptive to learning new skills. Like a sponge, a seven-week-old puppy absorbs everything you teach him the same way children do. He is responsive and reactive to all exposures. It is also during the first seven weeks that a puppy learns how to be a dog. He learns about rank and bite inhibition and starts on the eternal road of socialization. This is the best time to start his training and communication with humans.

When to Get a Puppy

If the puppy will not be left alone for more than three hours at a time, it is best to get a puppy that is about seven weeks old. He should not be separated from his mother or littermates prior to this. By seven weeks, the puppy has learned social skills from the other dogs but is still young enough to not be "damaged goods," not having acquired bad habits yet. He is still at an age when you can shape and mold him into a healthy, balanced dog.

Puppy Development

People's expectations of how long it takes a dog's training to advance to the next level are often unrealistic. Dogs are babies until they are about four months old. This is actually the easiest training period because they are easily guided. In fact, the window of opportunity to introduce novel stimuli and properly begin to socialize your dog is between five and twelve weeks. After six months, puberty hits and they become adolescents. Like an unruly teenager, they want it all, including the keys to the car. Obedience will be more of a struggle. Not to worry, your dog still loves you. They emerge from adolescence around eighteen months. Most breeds become mature at around two to three years of age. The smaller the breed, the faster the maturity.

The people who are eager to know when their dog will be perfect usually end training prematurely. Maturity takes time and patience. You need to accept your dog's age and developmental limitations. You would not expect a kindergartner to succeed in college. Please do not have similar unrealistic expectations about your dog.

The Personal Puppy Test

By the time I got Ruby I was deeply entrenched in the world of dog training and much more aware of how to pick out a puppy, or so I thought. I had read about performing puppy tests when choosing a new dog. Puppy tests are not written in stone but when performed at forty-nine days of age, they can give you some indication of the puppy's psyche as well as how he may react to different stimuli. It lets you know "who," to some degree, the puppy is—at least at that moment in time.

The puppy will probably have one of three responses: independence, insecurity, or interest. The independent puppy will probably ignore you when you test him. He will be more apt to do his own thing. He may or may not participate. Conversely, the insecure or fearful puppy will shudder in fear and will either run away, freeze, or growl at you. He may even be so afraid he will urinate when you perform the test. You want the puppy that is interested in you. An interested puppy displays a lot of social attraction. For this puppy, you are the draw.

Test each puppy individually so that you do not have to compete for attention with the other puppies. Initially, if faced with a litter of puppies, put them together and see what they do. The one that comes to you first is probably too bold. Don't take the one in the corner, either. He may be too shy. Test the remaining puppies separately. It is also important to test each puppy in a room where he has never been before. A new room provides novel stimuli to check if the dog is more interested in you or his environment.

It's a Gamble

These puppy tests should give you some indication as to what type of dog you're dealing with. I wouldn't bet the farm though. These tests are not always reliable. When I picked Ruby, I had a choice of three females from the litter. They all performed about the same on the puppy tests. I was frustrated and completely confused as to which puppy to choose. Two of the eight-week-old puppies began to fight so viciously, they drew blood from each other. Exasperated, I chose the one not fighting—Ruby.

Had I been thinking clearly and not been so emotionally overwhelmed, I would have realized Ruby was in fact the most aggressive, and because she was so tough, the other two were afraid to fight with her. No matter how friendly she is or how much she makes me laugh with her antics, Ruby is still an independent and dog-aggressive canine. Knowing this about Ruby, I've learned to work within her limitations.

Administering the Test

1. **Walk away from the puppy.**
   Put the puppy in the middle of the room and walk away. Does he follow you in blind faith or does he stand there totally stunned?

Does he sniff the ground? If so, he may be more interested in the ground than in you. Does he run off in the corner and hide? Watch how the puppy responds to his new environment.

2. Call the puppy.

   Once you are several feet away from the puppy, call to him, "puppy, puppy, puppy," in a beseeching voice. Each type of puppy will have his own response. The one that comes to you in leaps and bounds, thrilled at being summoned by a human, should rack up points. Does calling him even work, or do you need to clap and coax him to grab his attention?

3. Test noise sensitivity.

   Shake some keys to check the puppy's startle response. How does he respond to noise? Noise plays a part in training. You want to be able to manipulate a dog through sound. At the same time, you don't want a dog who is so reactive to noise that the slightest sound sends him flying across the room.

4. Check bite inhibition.

   Bite inhibition is the degree to which the dog bears down when he bites something like an arm of a chair or even an arm of a human. Offer the puppy a treat and see how he responds. Does he chomp down on the treat and your hand, or does he gently lick at the treat only? If you have a dog that does not learn proper bite inhibition, you could be faced with several future problems, including nasty lawsuits.

5. Do a target test.

   Try to get the puppy to look at your hand. Move your hand around, and check how easily the puppy follows it. Obviously this is important since training uses both verbal cues and hand signals. The easier it is for you to command the puppy's attention, the easier the training process.

6. **Perform elevation and restraint tests.**
   You want to see how the puppy responds to being handled. Lift the puppy under his chest so that his legs dangle off the ground. You can also hold him in a cradle position. With a very young puppy of four months or less you can perform a muzzle grab. Does he let you hold him or does he freak out? Some dogs are so overwhelmed and afraid, they urinate because you have done nothing more than hold and restrain them. Such puppies have not been handled much. It does not mean that they are bad dogs but that they've had a lack of social interaction.

The Muzzle Grab

I learned the muzzle grab from watching Lotte interact with puppies. When a puppy gets out of hand, Lotte places her mouth over the puppy's muzzle until the puppy becomes docile. While her action looks aggressive, she does not harm the puppy. She teaches the puppy that his behavior is unacceptable. The muzzle grab works brilliantly in the dog world and adapts well to human use. It gives you more control over your puppy, and when used successfully, it allows you to handle your puppy gently yet effectively.

Two warnings: the muzzle grab is not for the weak at heart and should only be used on a young puppy. Some puppies resist and thrash about while being held. It looks very violent and abusive when in reality it is not. It's like holding an out-of-control child until he calms. Some dogs may even be so startled they may dribble a bit of urine while you hold them. Be prepared for anything.

Muzzle grabs should only be performed on a seven- to sixteen-week-old puppy. At five months, the puppy's adult canine teeth begin to erupt. If he responds aggressively to a muzzle grab, you could end up with a bad bite from his larger, more powerful mouth beginning to fill with adult teeth. Also, the dog is almost an adolescent at that age and cannot be treated like a baby.

Seated muzzle grab

Performing a Muzzle Grab

1. While seated, hold the puppy in a cradle position in your lap with his back to your chest.
2. With one hand, grab the puppy's front legs at the chest so that you are holding him at his armpits.
3. Push the heel of your hand against his chest to stabilize the puppy against your own chest.
4. Wrap your free hand around the side of his muzzle and hold securely.
5. Very often, the puppy will flail, scream, fight, and carry on. Hold on even tighter.
6. When the puppy begins to relax, then you can relax your grip. Stroke the muzzle lovingly as you hold it.
7. Keep your hand tucked under his front legs and very slowly offer your other hand to the puppy by placing it in front of his eyes and mouth.
8. The puppy will either ignore, lick, or bite your hand. If he bites, he did not take the muzzle grab seriously and you must do it again, longer and possibly more firmly.
9. If at any time, the puppy starts thrashing again, go back to the pressure hold, and make it longer and stronger.
10. Release once the puppy is completely relaxed. He will probably take a deep breath and sigh before he concedes.
11. Try a treat test.

The goal is to find a puppy that either licks or ignores your hand when it's offered. That puppy shows submission to you. He may not be totally thrilled with you or he may be completely accepting of your actions, but he will stop struggling with you.

The Treat Test

The final part of the muzzle grab is the treat test. Whether or not the puppy will take a treat after the test gives you good insight as to the puppy's temperament. A puppy that takes treats is probably more confident and has a stable temperament. The other puppy, the one that will not take the treat, might be more stressed or unforgiving. He is probably too shell-shocked that someone stood up to him. It is as if he is saying, "I'm too upset and I couldn't possibly eat." Ultimately, you want the cool dog that goes along with what you say.

Common Errors

People have told me they tried a muzzle grab but it did not work. Most likely, it did not work because they did not do it right. Beware of three common errors people make when performing the muzzle grab:

1. Don't grab the muzzle without restraining the body. A puppy with all fours on the ground has a better chance of wriggling free.

2. Don't quickly grab the muzzle, squeeze it, then let go. This is not sufficient; it's like an antagonizing slap in the face. You need to hold the muzzle long and hard with a slow release.

3. Don't let go of the body restraint too soon. You need to continue holding the puppy's body against your chest and his paws in your hand after you release the muzzle and while you offer your hand.

A successful muzzle grab will help you read a bit into a puppy's temperament when you are trying to decide whether or not to add him to your family. It is also a good way to adjust your dog to the idea that you are in control. Finally, it is a good way to accustom your dog to being handled, which is important for grooming and medical purposes.

More Muzzle Madness

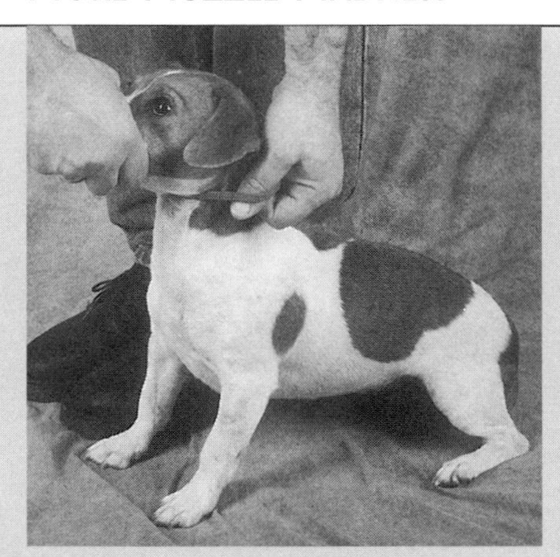

Standing muzzle grab

You can use the muzzle grab for more than just testing puppies. It also works well as a reprimand for a mouthy puppy. Let's say your puppy, tagging along at your feet as you walk through the house, suddenly grabs the edge of your skirt, pant leg, or shoelace in his mouth. Whip around, say "off," grab the puppy's collar with one hand as a substitute for the cradle position, and firmly wrap your other hand around his muzzle. When the puppy submits, offer your hand to check his reaction. If anchoring the puppy by grabbing his collar is not sufficient and the puppy continues to flail, then it will be necessary to perform a seated, cradle-position muzzle grab.

Initially, the muzzle grab may seem like a harsh way to handle your puppy, but it's not. Actress Laura Dern came to that conclusion when she saw the results of the muzzle grab. She called me to help with her ten-week-old mouthing terror of a puppy. I had never before met Laura or her mother, Diane Ladd, who was present when I arrived to find Laura's Labrador mix, Buddy Love, totally out of control.

I probably should have warned them before I rushed into action, but I bypassed the formalities and immediately switched to autopilot. I picked up Buddy Love in a cradle position and did a muzzle grab. Laura and Diane stared at me with their jaws dropped. Instantly, their expressions turned to gratitude when they saw how quickly Buddy Love responded and how he immediately transformed into a sweet, loving little guy. The muzzle grab is not about breaking spirit.

Shark Mouths

The inside of a puppy's mouth is filled with sharp, tiny milk teeth. It's hard to believe sometimes how much pain a puppy can cause when he bites you with those pearly whites. The puppy is so cute and his teeth are so small, you wonder how he could deliver so much pain with one little bite. A puppy that sends you flying around the room shaking your hand to relieve the pain after he nips you has no bite inhibition.

Bite Inhibition

A puppy needs sharp teeth to compensate for a weak jaw. It is how he is able to cut through and eat meat, a sinuous substance. Also, sharp teeth help the puppy learn about bite inhibition. A puppy learns a little bit of pressure causes pain for his victim. By the time the puppy grows up and has real teeth and a strong jaw, he learns how to control his bite. This is obviously important because a grown dog can do quite a bit of bite damage given his paraphernalia. Puppies should learn the majority of bite inhibition between five and seven weeks of age.

A puppy hones bite inhibition when he plays with his littermates. Here's how it works. Littermates Suzy and Al are playing. When Al bites Suzy, she yelps and stops playing with him. All fun ends and Al is shunned. Al tries to wrestle with Suzy again later and bites her once more. This time, not as hard. She continues to play with him, teaching him that a softer bite does not hurt as much and is acceptable. If he bites too hard again, she stops playing. The puppy learns through trial and error that if he bites too hard, no one will play with him.

The mother also teaches her puppies about bite inhibition through play and during nursing. Lotte would have been a great mother. If a puppy ever got out of line or bit her teat, she would have been right on top of it. When Lotte is around overly pushy puppies, she performs the quintessential muzzle grab. She pins the puppy to the ground with her mouth over the puppy's muzzle. To the novice, it is frightening to watch because Lotte has a long muzzle herself and she exposes her fangs when she grabs the puppy. She gets her point across without cutting, causing damage, or even so much as shaving a hair off the puppy. Nothing. Lotte has bite inhibition down to a science.

Teaching Bite Inhibition

When you offer your puppy a treat, he may lick at it and gently place his teeth on you. Once in a while a puppy may nip at your hand. Some puppies are all teeth. They really crunch down, not understanding how to hone in on just the treat, and take your hand as well as the treat. These puppies may have been orphaned or removed from their littermates too early, or they're just "that way." Obviously, they require more work.

If you have a puppy that is only five weeks old and he marks you with his little shark mouth when you give him a treat, don't be surprised. He is a very young puppy that has not had those couple of weeks of socialization with his mother and littermates. He may tend to be extremely mouthy and have a problem with bite inhibition. Teaching bite inhibition to this puppy is now your responsibility. You take on the role of his mom and littermates. First, try a muzzle grab on an age-appropriate puppy. Two other methods are the "ouch" and time-outs.

The "Ouch"

There is a technique referred to as the "ouch." When a really young puppy bites, you should pull away and yell, "OUCH!" In theory, the puppy associates his action of biting with your reaction of pain. He will probably test the water a couple of times. Each time he bites, respond with "OUCH!" This works well with very young, sensitive puppies. Although this is a very viable technique, it is not problem-free. Sometimes it backfires. When you scream "OUCH!" you may excite the puppy. He thinks the sound you made is cool and wants to play more. With an excitable, pushy, aggressive puppy, the "ouch" could actually worsen the situation.

Time-Outs

When a muzzle grab and the "ouch" are ineffective because they both inflame the puppy's aggression, an alternative is a time-out. When the puppy is in a biting mood, put him in his crate and wait until the mood subsides. Please don't assume he is thinking about what he has done wrong or feeling remorse about his situation. You are simply breaking the behavior pattern. However, he may realize that biting leads to social isolation if repeated. For this reason, the time-out need not take more than a couple of minutes to have impact, unless you are in need of some "crate rest."

Bite inhibition is a slow process since you are teaching the dog to control and restrain his jaws. A dog that never properly learns this can be dangerous. I receive numerous frantic phone calls from people whose dogs have bitten someone. The dog that inflicts multiple bites and draws blood, possibly even requiring sutures, is a dog without bite inhibition. Not only did he bite once, but he went back for more. Luckily, that's pretty rare. More commonly, I see dogs that never mouthed as puppies. They never learned the power of their jaws, and the one time they bite someone, their teeth cut through skin like butter. People tolerate many behavior problems. Biting is not generally accepted and is often the reason a dog is euthanized.

Parenting Your Puppy

I hope you're beginning to understand the importance of a puppy's first few months. Much of how your puppy deals with the world depends on how he spends his young life. When you bring home a puppy, you bring home a living being that depends on you. Raising a puppy requires a tremendous amount of responsibility. We would never leave a child unsupervised while we go off to work, yet some people think nothing of leaving a puppy alone for eight to twelve hours a day. How you begin your relationship with your puppy lays the foundation for your future life with him. You need to be with your puppy as much as possible to guide him through our world. Remember, dogs are not humans. How they should relate to us and our world needs to be thoroughly communicated to them.

In Short

The ideal time to bring home your new puppy is at seven weeks of age. By this time the puppy's brain is fully formed, and therefore capable of being trained. He has been with his mother and littermates long enough to begin to learn bite inhibition, rank, and how to respond like a dog. He is also still young enough to easily merge into a new family pack and to begin socialization.

When choosing a young new companion, puppy tests can be a helpful and insightful tool. In order to get the most information possible, first observe the litter all together to see how a puppy responds to you and the other puppies when you present yourself. Then move to an area that is new to the puppies and begin testing them one at a time.

To administer the puppy test:

1. Walk away from the puppy to see if he is interested in you.
2. Call the puppy to see how responsive he is to verbal signals.
3. Test the puppy's noise sensitivity by jingling some keys.
4. Check bite inhibition by offering him a treat.
5. Test his targeting ability by moving your hand and observing whether or not he follows the movement.
6. Perform elevation and restraint tests to examine the puppy's response to handling.

Is the puppy independent, insecure, or interested? You want the puppy that is interested in you. While these tests are not the be-all and end-all of what these young dogs are or will become, they do give you information on which to base your selection.

When your puppy needs an attitude adjustment, perform a muzzle grab. When used successfully, it allows you to handle the puppy gently yet effectively, calming him down. A muzzle grab should only be performed on puppies that are seven to twenty weeks old.

One of the biggest problems new owners face with their new charges is puppy mouthing. When a canine youngster throws a tantrum by using his needle-sharp milk teeth, the best ways to get him back under control are through either the muzzle grab, the "ouch," or the time-out. Each has its uses based on the age and temperament of the puppy and the intensity of the tantrum.

Part II
Relationship

Building a Lifelong Partnership

MONOPOLE
LA TACHE

4

What's Essential to Dogs

Our biggest communication problems with dogs occur when we punish them for doing what they are supposed to be doing. Dogs chew, chase, sniff, bark, tug, fight, play, eat, sleep, and dig. These are things they are supposed to do as dogs. It is our job to give them appropriate outlets for their natural behaviors. Don't punish your dog for being a dog. Instead, teach him how to live within the structure of your life. For instance, the dog must learn not to respond to his innate predatory drives when he is attracted by the fluttering of a bird or the wheels of a bicycle, but he can herd a basketball around the backyard or play with his friends at the dog park.

Pack Animals

Dogs are social animals. Beginning with their earliest ancestors, dogs have traveled, eaten, slept, and lived in packs. It is how they survive. Consequently, they think of themselves not just as individuals but as members of the pack. This is one of the reasons we get along so well.

Dogs instinctively view us as part of their pack. The pack may be as simple as a single bachelor or as large as a family of ten. You and your family make up your dog's pack.

To a dog, the pack equals survival. Without a pack, some dogs cannot survive emotionally; the separation anxiety is too severe. The pack is so important that when a dog is separated from the pack, he will do anything he can to rejoin his clan. For example, if a dog is stuck in a ditch he howls, barks, and cries for the pack to return and rescue him. The same thing happens when you leave your house. He is calling for your return. Of course, many dogs are fine when left alone. It all depends on the dog's personality and life experience.

For humans, being an individual is very important because it symbolizes self-sufficiency and strength. It means just the opposite to a dog. Dogs have a powerful life force which radiates in a pack. We must take into account the fact that dogs work best in a very rich and organized pack society.

Family-Oriented Animals

A dog's pack is his family. It is his clan or tribe. Hence, members build strong bonds with one another. They form lifelong relationships. Just like we have roles in our families—mother, father, daughter, son, niece, nephew—so do dogs. Each dog has his place within the pack: some are hunters, some are protectors, some are lucky enough to be top dog, and some are the runts. Each dog has his own position in the family. Sometimes a dog's position can change. Depending on circumstances, a low-ranking member can be elevated to a higher level.

Pack Hierarchy

We often refer to pack hierarchy as rank. Within the family unit, dogs have a hierarchy. Without it, pack life would be a free-for-all, resulting in injuries. There is a strong leader with subordinates. Dogs vary from being king to being second in command to being third, fourth, fifth, and so on. You might know these levels as alpha, beta, all the way down to omega. The omega is usually the patsy that takes the brunt of everything. It's his lot in life, and we should not feel sorry for him.

Sometimes these dogs rise in rank. There is really no such thing as an absolutely submissive dog or an absolutely dominant dog. You can perform all of the puppy tests and dog tests in the world when deciding on a dog, and ultimately, they do not mean a thing. Rank is fluid; it is situational, conforming to the social interaction of the moment. For example, you take your dog to the dog park every day where she interacts with her friends. These dogs know each other and have established a hierarchical order. One day, a new dog comes to the park. The dogs reestablish pecking order to find a place for the new kid on the block.

All dogs need a pack with a strong leader. That is one of the reasons they fit into our lives. Dogs have an allegiance to us as their family pack. They feel safe with us when we assume the leader role. Problems occur when we create fuzzy boundaries or inconsistent rules for them. For those who think letting your dog make most of the decisions for himself is being kind, think again. It confuses your companion and will have behavioral repercussions. Being the top dog does not make you mean. It gives you the leverage you need to allow your dog to feel safe and secure about his survival.

Adaptable Creatures

Canids are highly adaptable. Coyotes, for example, still live in areas where 99 percent of their territory has been taken over. Canids are very good at making due. That is what is so fabulous about them. It is why we can bring dogs into our homes. They adjust to our surroundings as long as we set rules and teach guidelines to ensure a healthy adaptation. If a dog's needs are not met and you feed into its aggressive and predatory instincts, then you run into some major problems.

Procreation

Obviously, one thing that ensures the continuation of any species is bearing young and then caring for them. This is most likely why intact males want to mate with anything and everything they see, including lampposts and human legs. Theoretically, mating is limited to the two top dogs in the pack because they have the strongest genes to pass on. There is also only one litter per year to keep the pack down to a reasonable number. While it is good to be part of a large pack, it is also important that the pack not get too large, or there might not be enough food to go around. Domesticated dogs do not need to worry about finding food since we give it to them. They can afford more than one estrous cycle per year. This is one of the ramifications of domestication.

Real Men Neuter Their Dogs

It is a more natural state for a dog to be neutered. A dog in the wild will most likely never mate. If you really want your dog to be natural, then spay or neuter it. We have created a dog's artificial, constant desire to breed.

Don't personalize neutering your dog. Actor James Caan had an Akita mix named Reno. He was bent on keeping Reno intact. Because of several problems James had with Reno, I suggested neutering as his salvation. Luckily, his wife sided with me, and we were able to sway him to neuter Reno. James felt it was more natural to keep his dog intact. This is not an uncommon feeling among men and sometimes even women.

I've encountered a few people whose feelings on the subject are so extreme that they cannot bear to part with their dog's parts. In these instances, I have suggested "neuticles," the actual addition of artificial testicles. As unnecessary as this may sound, at least the behavioral benefits of neutering have been achieved. Canine testicular prosthesis is the doggie equivalent of a Hollywood boob job.

Dogs as Predators

Wild canids are predators that use their wiles and have family and friends to help them satisfy their needs. They are meat-eaters, which means they kill. When you get a dog, you bring a predator into your home.

Look at dogs for what they are. Luckily, thousands of years of domestication have greatly arrested the killer instinct. Most dogs will never "kill" anything other than a box of crackers. There are similarities between humans and dogs, but realize these animals have the equipment to do a tremendous amount of harm.

Most dogs that bite are not trying to kill. Rather, they have been put in a situation where they feel they need to use their main defense mechanism: their mouths. Many people are unaware of this fact when they leave a dog alone with their children, and then they blame the dog when they need to rush the child to the hospital for stitches.

In wild canid packs, the chase, the kill, and the feast were all highly ritualized behaviors, the echoes of which are still innate in today's domestic dogs in varying degrees. Since we feed our dogs, they do not have to hunt. We have domesticated the killer instinct out of the dog. But even though we provide food for our dogs, their hunting instinct is still intact. When your dog sees a squirrel or a bird, he may assume the classic hunting postures and behaviors. He stalks and pounces but rarely makes the mark.

Our problems with dogs arise when we fail to see that they are predators. We take for granted their ability to assimilate into our pack structure, which does not include predatory behavior. When our two worlds collide, we chastise and punish our dogs for our own lack of insight. We do share some traits with dogs, but we wrongfully assume that they share ALL our traits. By doing so, we ignorantly disregard what they are, and then we punish them for being that way, rather than working around and with it.

Prey Drive

Dogs are absolutely attracted to things that move. It is not unusual for a dog to lunge after a moving car, skateboard, baby buggy, truck, or jogger. Movement naturally attracts dogs because of their prey drive. Dogs are also attracted to high-pitched staccato sounds similar to those made by a wounded animal. When dogs see a squirrel running through the park or hear a child squeal, they go into prey drive. In this situation, a dog treat may not be enough to attract your dog. Unless you have a strong relationship and a history of conditioned responses, the dog may completely tune you out.

Displays: How to Read Your Dog

Dogs go through many ritualistic displays of dominance and submission due to a strong inhibition against hurting another animal in their pack. What has evolved from their bite inhibition are these great behavioral displays: snarling, growling, posturing. The dog uses any means necessary to get his point across without the actual use of deadly force.

A dog's body posture and facial expressions tell a lot about his state of mind. Dogs move either in a forward or backward direction, making their bodies appear larger or smaller. A confident dog may show his dominance and superiority by moving forward with a larger stature, trying to make himself appear bigger; a submissive or fearful dog tends to lean backward, minimizing his overall posture.

Body Postures

Levels of Dominance

Aroused: slightly leans body forward; slightly elevates tail; focuses eyes ahead; bends ears forward.

Dominant: generally stiffens body and legs; raises tail higher and possibly begins to wag stiffly; raises hackles.

Aggressive: totally stiffens body and legs; raises tail straight up and may wag stiffly; focuses in a direct stare; snarls and possibly growls.

Dominant-aggressive dog snarling and holding ears forward

Dominant-aggressive dog stiffening body and baring teeth

Levels of Submission

Moderately submissive dog trying to look smaller with a lowered, crouched body posture

Early submission: slightly leans body backward; focuses eyes softly; slightly folds back ears.

Moderate submission: crouches body back; begins to tuck tail; avoids eye contact; folds ears back; may curl lips back in a "horizontal grin."

Advanced submission: lowers body to the ground; completely tucks tail, covering all sources of scent.

Extreme submission: may possibly turn on side and lift leg to expose belly; may occasionally dribble a bit of urine.

Levels of Fear

Fear-aggressive dog flattening ears backward and baring teeth

Early fear: crouches body position; tucks tail; avoids eye contact; folds ears back; possibly shakes; shows overall avoidance.

Fear aggressive: lowers body and leans back crouched; tucks tail; flattens ears backward; snarls; growls.

Dogs are masters of body language. Body postures let a dog know how another dog feels about a given situation. Our dogs give us many clear signs of how they feel. Yet to the novice owner, these displays go completely unseen. This is why many people are surprised when their dog adds some bite to his bark. I hear far too frequently, "He bit me out of the blue." Chances are, if the owner had been a bit more aware of his dog's behavior, he would have realized that the dog sent some pretty clear warning signals before lashing out.

Healthy Outlets

It can be frightening to see your dog display predatory behaviors when he shakes a toy, growls, or snarls. People are often frightened and dismayed at how voraciously a dog attacks something as benign as a toy. The behaviors of violently shaking and gripping are all part of being a predator. This is something owners need to come to terms with and understand as normal dog behavior.

We should not be frightened or, conversely, take it lightly when we see our dogs make these displays. We need to understand that, in its ancestral context, this display could mean death. If your dog is shaking a toy, let him. It is a natural behavior. You are giving your dog a healthy outlet for a normal predatory behavior. You don't want to stifle these behaviors because eventually they will show up in other ways. Drain the behavior in a positive way before it surfaces in a negative form.

Boiling Point

Even the tamest of dogs has his boiling point, and his suppressed drives may eventually burst upon another animal, a child, or even you. Think of it as a pot of oatmeal. You can put a lid on the pot to keep the oatmeal contained. As the oatmeal cooks and expands, it needs to find some way to get out of the pot. It can bubble up and push its way out, seeping through the crack between the lid and the pot. It can even blow the lid off all together. The point is, you cannot completely contain it, just the way you cannot stifle a dog's natural instincts.

You must give your dog some sort of outlet for his natural predatory instincts. Maybe you can put a lid on his chewing behavior, but that energy needs to come out somewhere and it just might appear in the form of excess digging. If you cap digging, that energy might bubble out in the form of chasing. Then again, if a behavior is suppressed too long, it might very well blow the lid off all together, sticking you with an awful mess. You cannot stifle these natural behaviors. They will come out in different ways.

No Dog Is Exempt

You may not be able to picture a cute little Cocker Spaniel as a predator. But do not take what I say lightly. All dogs have a predatory instinct. It may be suppressed, but it is there. Fortunately, they don't try, or even want, to hunt us down and kill us. However, when they feel threatened or trapped, dogs may use their main mechanism of defense: their mouths. People may not worry about dogs of a smaller stature because we physically outsize them. Can they kill you? No. Can they scar you? Yes. Any dog can send you or your child to the doctor for stitches.

Territory

Some dogs defend anything they perceive as theirs. That could mean your house and yard, you, bones, other dogs, females, young puppies, and themselves. This is one of the reasons we love having dogs. They protect us from potential harm. My dogs give me a sense of security. For some people, having a dog for protection runs neck and neck with having one for a companion. Initially, I got Ruby to act as a sentry. I wanted a menacing-looking dog. When I learned that the Dalai Lama used eight Tibetan Mastiffs to guard his summer home, I romanticized the idea of having one of these same dogs standing post at my new home.

In the larger scheme of things, protection of territory means survival. Whoever controlled the largest territory containing the largest amount of prey was going to be the most successful in terms of survival. This is why some modern-day dogs define their territories by marking and why some bark when intruders approach. Dogs have an entire repertoire of behaviors, including barking, growling, lunging, and biting, which they use to defend what they perceive as theirs.

Marking

Dogs use scent to mark their territory—hence, the behaviors of urine-marking and scraping after defecation to call attention to their feces. Whether your dog is intact or neutered, don't let him mark incessantly.

It can be a precursor to aggressive behavior. He is saying, "I am male. Let me mark." That dog may think he owns the area. He is leaving his calling card saying he was the last one there, which reaffirms he's the one in charge. If you have a dog that tunes you out in order to mark, don't let him mark. Keep him on a leash and moving forward.

Scent

A dog's sense of smell is much more acute than ours. Olfaction is one of a dog's primary methods of communication. The world of dogs is much different from ours. We can't even begin to understand what it's like to live in a world so profoundly ruled by scent. The most we notice is someone who wears too much perfume or has bad breath. Dogs read volumes in what they pick up through scent: they learn about other animals or beings, they track, they give information about themselves. Imagine the myriad of scents your dog picks up on a sidewalk. And he is supposed to listen and come to you no matter what? Yeah, right.

Unfamiliar Dogs and Their Scent

When two unfamiliar dogs meet, the most normal greeting is a nose-to-butt sniffing fest. A lot of information is gained by sniffing: sex, reproductive status, age, health, and we don't even know what else. It's important to let your dog sniff another dog. Make sure your dog is on a loose leash when he goes in for a sniff. A tight leash signals the dog to get uptight. It warns him of impending doom. When two well-socialized dogs meet and a fight ensues, it is usually owner induced. The owner pulls too hard on the leash or gets nervous.

Allow your dog to sniff other dogs. It is bad dog etiquette not to take part in the sniffing ritual. And it *is* a ritual. One dog stands and allows a sniff, then the other stands and allows a sniff. If either dog becomes jumpy or pulls away, he may incur a retaliation by the other. A dog that is not properly socialized usually fails the sniffing ritual. The other dog will either ignore him or become aggressive. This is also where the owner's part is important. The owner must also be socialized to dog-sniffing behaviors and allow the dog to sniff. A well-socialized dog is happy to go in for a quick sniff and then continue on his merry way. Hence, the incredibly important socialization process. If you want to pass dogs on the street, that's okay, too. Just make sure you give your dog other opportunities to say "hello."

Play

Some experts have posited that dogs are like juvenile wolves. Their development was arrested in a sort of limbo state of adolescence. The more serious behaviors—the true killing behaviors—while still intact, were never fully developed in dogs. They are perpetual puppies that never matured to the point of the wolf. Being perpetual puppies, they are far more vocal and playful than wolves.

Play incorporates learning skills, bite inhibition, social skills, exercise, and entertainment. Simply stated, play is fun. The purpose of play is to mock hunting skills. It involves ducking, bowing, grabbing, wrestling, and other tactics that help reinforce hunting behaviors. It also helps reaffirm close bonds.

A Part of Life

Play is an essential part of a dog's life that you need to encourage. Not only is it fun and good exercise, it is also an excellent social outlet. Your dog must learn how to play properly. Dogs continue to play all their lives. Puppies learn life skills through interactive play. They learn what they can and cannot do. If they get too rough, they get in trouble.

While play is a puppy's rehearsal for real life, to an older dog it's just fun. A dog that does not know how to play is the saddest thing I've ever seen because he never learned how to interact with other beings, especially his own species. It is your job to properly socialize your dog so that he learns these skills.

Pure Communication

I take Ruby and Lotte for a hike every morning, and I love to watch Lotte seduce either another dog or a person into playing with her. She picks up a stick and shows it to her target. It dangles from her mouth like a cigar.

She looks at the person as if to say, "Do you like my stick? Don't you want my stick?" She's hoping to engage them in a bout of chase-me-to-get-my-stick. Besides eating, seducing dogs to run after her to get the stick is Lotte's favorite thing to do. It is an absolutely heartwarming and delightful process to watch. She's not thinking of hunting squirrels. She's communicating. I always baptize the stick by naming it "dog toy" or "chew toy." When I say those words to Lotte, she grins with ecstasy and begins prancing around and tossing the stick in the air.

What Is Gained by Play?

Play teaches a dog bite inhibition. Through playing with his littermates, a puppy learns the power of his jaw. The other puppies won't play with him if he bites too hard. He learns rather quickly that if he's gentle, he gets what he wants: more play. This is part of learning how to socially interact with another being. Play teaches the dog how to participate in the dance of life. He learns when to back down or be assertive.

There are dogs I call good players. Lotte is an eloquent player. She speaks dog fluently. So did Pearl. While Ruby plays well, she is a big oaf. She knows the rules of the game, but she doesn't know her own strength. She needs to play with a dog her size. Different breeds play in different manners. Play generally mimics the breed's original function. Herders prefer a good, old-fashioned game of chase whereas the smaller of the terriers prefer a scrappy wrestling match. Body slamming is a favorite among bull terriers, pit bulls, and other large working breeds. This is why you need to monitor your dog's play sessions. Injuries may occur when dogs of breeds with different playing styles get together.

The Play Bow

Posturing during play is of the utmost importance. The play bow is an invitation to play and an indication to the other dog that he should not feel threatened. The dog's front paws and head are low to the ground and

A play bow is an invitation to play.

his butt is up in the air so that his back is bowed down toward his forepaws. In dog language, the play bow means everything that happens after the play bow is done in jest. If the dog growls, in this context it should be perceived as a play growl; if he mouths, it should not be perceived as a threat but should be received as fun. The play bow tells another dog to come forth, play, and interact in a friendly game of tag, chase, wrestling, tug-of-war, or catch-me-if-you-can. Be warned that sometimes a play session can become more aggressive, like a friendly conversation can become an argument.

Mouthing

A puppy explores his world via his mouth. Think of babies and toddlers who find new objects. The first thing they do is stick the object in their mouths. Puppies are also exploring their relationships with their owners or other humans. They are testing to see what they can get away with. If the puppy were still with his littermates, he'd bite them to have a show of strength. They bite to see how far they can go. When the littermates are no longer present, puppies bite other members of their pack: you.

Puppies test the waters with their teeth. They want to know how much they can bite you and how much you will tolerate. That's where you have to treat them like another dog would. I went into one household where the puppy was mouthing everyone. The resident dog nailed the puppy with a muzzle grab. You should take your cues from your other dogs. Why is it okay for a dog to reprimand another dog, but it's not okay for us? It is not cruel; it is how dogs learn. Being a strong teacher fortifies your relationship with your dog.

Chewing

There is a period in a puppy's life when chewing helps relieve the pain he feels when teething. However, people tend to make excuses for either severe, destructive behavior or heavy mouthing and biting by saying, "Oh, he's teething." That may be part of it, but the teething period is short-lived. The truth is that dogs need to chew throughout their lives.

They need an outlet for their normal chewing behavior. You must let your dog chew on appropriate chew toys.

Chewing keeps a dog's jaw strong, and it's good for his teeth and gums. Dogs also chew because they don't have opposable thumbs. In order to investigate, a dog uses his mouth because it is the body part with the most dexterity. Besides, think of the many bad things your dog can't be doing if he's chewing an appropriate chew toy. It makes an excellent tension, energy, and stress reliever. After a good round of chewing on an appropriate chew toy, your dog will probably be so relaxed, he'll flop down and go to sleep. It's like getting a good massage.

Leave the Ego at the Door

We approach our dogs very egotistically by disregarding and discounting their rich, behavioral repertoire. Dog behaviors define what they are as creatures living on this planet for eons. We march into their lives and take over by dictating what they should do, how they should act, and what they should value. In the process, we ignore and stifle their specific, natural behaviors. The same process occurs whenever a people or a nation is conquered. The conqueror imposes his beliefs on the people and disregards their rich cultures. The conquered are always the ones to suffer.

Whether we rule over a nation of people or our dogs, something is lost when everything should be gained. We have a lot to learn from our dogs. It's a shame not to recognize this. A multitude of dogs are put to sleep because of our arrogance and ignorance. They pay for our failure to accept their differences. We tend to oppress what we don't understand. Once you realize what is essential to your dog, you will better understand his behavior and help him cope in the human world.

In Short

Pet owners run into problems with their dogs when they fail to realize they are taking in an animal capable of predation. There are certain behaviors such as chasing, tracking, and chewing that are innate to dogs. Don't punish your dog for being a dog.

Dogs are social creatures that function well within a family structure—whether it be human, canine, or a little of both. If we keep this in mind, much of how our dogs respond to the world around them makes infinite sense. Within their family unit, dogs look for a leader. Rank maintains pack harmony.

Dogs evolved from wild predators. Although years of domestication have arrested their killer instinct, dogs still have the equipment for potential harm. Luckily, dogs have many vocal and visual displays of snarling, growling, barking, and posturing that are used to make a point without chancing injury. Pet owners need to read such warnings to help prevent attacks or bites. They also need to provide outlets for their dogs' natural canine behaviors.

Play is a great way to foster communication between you and your dog. It incorporates learning skills, bite inhibition, social skills, exercise, and entertainment.

Chewing is the canine national pastime. Appropriate chew toys must be made available to occupy bored, underexercised dogs. If you fail to do so, you will pay with your furnishings.

Realizing what is essential to your dog helps you better understand his behavior. You can then help him assimilate into the human world.

5

Building a Leader–Follower Relationship

Dog training has two components: the relationship you and your dog develop and the nuts and bolts of training. Once your dog understands what you want, he must have enough allegiance to you to do it in all situations. Allegiance is the ultimate outcome of creating a leader-follower relationship. Dogs are either leaders or followers. In order to have a successful, healthy relationship with a human, a dog must be the follower. You can capitalize on what is essential to dogs and use this information to become a stronger leader.

Rank

Dogs view most everything in terms of rank relevant to themselves. There are certain rules, such as eating rights, sleeping rights, and entry protocol, that define and pertain to a dog's rank. Most people don't realize rank pecking order is interchangeable and situational.

Rank is never set in stone. For example, a couple has a dog. When the man is home, the dog views him as the alpha dog and the woman as the beta dog. They are number one and number two in the household. The dog is good and compliant when the man is at home, but as soon as the man leaves the house, the dog starts to act up. He bites and play-fights with the woman. When the man is not at home, the number one, or alpha position, is vacated. Number two must assume the number one role before the dog does.

This scenario translates to a variety of situations. Sometimes the dog views another dog as the dominant figure. In the wild, a lower-ranked canine can move up the ladder and assume the alpha position. Make sure you stay on top. Your dog must understand in no uncertain terms that you are his leader. Otherwise, you could end up with a dog that takes over your home. Your dog shines when he knows who is in charge. Dogs crave rules and boundaries set by a strong leader. So much is gained with a healthy leader-follower relationship, and so much is lost when a dog is left alone to do his own thing.

All in the Family

When there are two or more members in a household, anyone who wants the family dog to listen needs to be involved in training. Dogs tend to respond to household members who spend the most time caring for and training the dog. Many people falsely assume that a trained dog will follow any command given by any person. Because so much of training involves relationship, a dog responds better to people he knows. Just like children may not listen to a new baby-sitter or a substitute teacher, a dog may not listen to a person with whom he is not familiar.

Rules of Rank

Certain guidelines involving body language and tone of voice help ensure your number one position. By standing tall, using a lower-pitched voice with an authoritative tone, and making direct eye contact, you tell your dog you are a strong leader. The rules of rank help instill a powerful impression of you as the alpha leader in your dog's mind.

A Note About Rank

After you establish a clear hierarchy and there are no other behavioral problems, you can fashion rules about rank to fit your lifestyle. I am the leader in my house, and my dogs know it. I don't feel I dominate my dogs, but at the same time, I have the last word. There are also times when I allow them a little leeway because I find it amusing. I have raised them in such a way that challenging me at a core level is not an issue.

I love to have my dogs on the bed with me at night while I watch TV. When I tell Ruby to get off the bed, she'll sort of look up at me with her big lazy eyes as if to say, "Are you for real? Do I really have to get off the bed?" I'm amused by her response, and I understand the way she feels. If she were to growl at me or show her teeth when I tell her to get off, then I'd have a problem. But she doesn't. I tell her "off" a second time and she complies. I find Ruby's response endearing because that's Ruby. It's a part of her personality. I don't want a robot dog with a broken spirit. Rank is a very complex issue because you must have the final say, but at the same time, you want the dog to have his own personality.

Follow these five guidelines from the first day you bring home your dog or puppy:

1. **Sleeping Rights**

   The pack leader always has the best resting area. In the dog world, a higher sleeping area provides a better vantage point to scope out danger and prey. Generally speaking, to be physically higher connotes a power position; therefore, the more dominant dog always has access to the highest, most comfortable resting area. In the home this means the bed, couch, or comfortable chairs.

   With this in mind, if you allow your dog on the bed with you, you may be telling him he is equal in status to you. If he is on the couch with you and you pet him, it not only gives him a message of equality, but you are also giving him your undivided attention, which tells him he is like a god. You might as well be fanning him with peacock feathers, feeding him peeled grapes, and shouting "Hail Caesar!"

   While we're on the subject of royalty, make the dog aware that his bed is not his throne by moving it or occasionally sitting on it yourself. Again, if you do not have other behavioral problems with your dog, there is nothing more fabulous than snuggling your dog in bed.

2. Eating Rights

To dogs, food is a way of maintaining the pack hierarchy. That's why it's important to make sure your dog works for his food. Nothing is free. Give him a command like "sit" or "down" before you feed him snacks or meals. Not only does this help teach your dog good manners, but it helps reinforce commands. In the dog world, the one that eats first is the leader. If you are having a difficult time with your dog and he really challenges you, you might try adhering to eating rights more closely.

Feed your dog either several hours prior to your own meal or sometime thereafter. Don't fret if your family does not eat at a regular hour. It's more important that your dog work for his food. If you snack throughout the day, let your dog know you control the food and don't offer him anything.

Finally, do not free-feed your dog. If he can get to food all day, he will not make a connection between his owner and his food. Also, he never has a chance to get hungry, and hunger yields the secretion of gastric fluids, which help with digestion. Besides, dog food, both dry and canned, can become rancid and spoil when left out.

3. Earning Praise

When a dog nudges his owner's hand, he commands his owner to give him attention. The urge to pet or play with your dog may be strong when he does this, but resist the temptation and maintain your dominant position. This is not to say you should never pet or play with your dog. You can easily maintain your rank by telling him "sit" or "down" when he nudges you. Once he complies, you may pet him. Nothing should ever be free if you live with a pushy dog!

If you are not having any problems with your dog, it's fine to give him attention when he asks. We often unconsciously begin to pet the dog. In many cases a free pat is meaningless. However, a dog that has problems like noncompliance and aggression may subtly manipulate you when he nudges you. The dog says, "Hey, Owner, give me attention," and we comply like good little humans. The roles are clearly reversed. Some aggressive dogs get their owners to pet them, and then they walk away or growl. They test you by commanding you to start and stop. Then again, some dogs are just needy and love to cuddle.

4. **Games**

Games are a great way to build your dog's confidence and fortify your relationship with him. However, you must remember that the toys are your property and not your dog's. You own the toys, you control the games, you are the leader. You initiate the game, and when finished playing, you end the game and *you* put the toy away. That's right. You put it away. Do not allow your dog to run off with the toy or leave it around the house for the dog to pick up freely. Otherwise, he thinks he is the winner and therefore the leader.

Again, this is not cut-and-dry. If you are not having leadership problems with your dog, you don't need to be so concerned with this. I encourage Lotte to bring a toy over to play. It's so rare for Lotte to bring me a toy that, when she does, I know it's because she's in a playful mood. I'm not concerned that she feels she is in control. I'm rapt with joy when she wants to engage me in play.

5. **Entry Protocol**

Doorways may not have much meaning to you or me, but to dogs, they can hold great significance. The leader passes first through a door. Stay in control by putting your dog in a sit-stay before you walk through a doorway. This also helps enforce boundary training for the dog by slimming the chances of him bounding out the door and into the street where he could be hit by a car.

I have seen only a few dogs that really challenge at this level and try to beat you through the doorway. Again, if your dog doesn't have any other behavioral problems, you need not obsess over entry protocol. I step over Ruby when she lies in the pathway. She's not thinking of blocking me. She's just comfortable. My stepping over her does not create a problem. If your dog aggressively controls passageways by growling or challenging you when you try to walk through, then you need to solidify rank and fortify entry protocol. By dethroning "the king" you reclaim your leadership position.

Politically Correct Games

Be cautious of games like tug-of-war, wrestling, and chase because they are games of strength, power, and control. These games may often elicit unwanted aggression and behavioral problems because they teach the dog to view you like another dog, and oftentimes a lower-ranking dog.

Fetch is always a great game to play. To encourage your dog to play fetch, choose one of his favorite toys and make it special by keeping it away from him. Only bring it out when you want to play fetch. Tantalize your dog with the toy. When he is excited over it, toss it a short distance in front of him and say "fetch" or "take it." When he picks up the toy, encourage him to return to you.

Exchange the toy for either another toy or a treat and say "give." Immediately throw the toy again. Never give your dog any time to think you have taken away his toy. If you do, he may not want to relinquish it in the future. Stop playing fetch before the dog is bored. Always end with him wanting more. Slowly progress, throwing the ball or toy further. For most dogs, the act of retrieving becomes enough of a reward, and you can easily phase out the treat.

You can play a politically correct version of tug-of-war with a nonaggressive dog. It's actually a good confidence booster for a shy or underconfident dog. Certain rules apply: you begin and end the game, and the dog must release the toy upon command. A give command becomes a control switch. If your dog is confident, make sure you win each match. With an underconfident dog, you let him win and then you win. Balance the victories. Once his confidence rises, you win each match.

Remember to put the toy away at the end of every play session, reinforcing your possession of the toys. This will help establish your leadership role.

Nothing Is Free

Initially you want to make sure your dog works for attention and praise. Even though your dog or puppy may be the cutest darned thing around, try not to succumb to his wishes and dole out free attention. It only takes one extra second to tell your dog to sit before you pet him rather than just automatically reaching down and petting him for free. This is particularly important when you first start training your dog. Don't even give him a chew toy or take him for a walk without making him sit first. Those should be viewed as privileges by your dog. This is an easy way to maintain rank, but it is also not permanent. Once your dog figures out you are on top, you can ease up and hand out a free pat every now and then.

Enforce All Commands

Do not give a command unless you can follow it through. Once you give a command, you are committed to it. Don't just throw it out casually and then concede when the dog doesn't comply. The dog will view you as inconsistent and not worthy of response. When you're lying on the couch and you call your dog to come, you'd better make sure he does. Don't dismiss the command if your dog does not come over to you. Instead, get up and gently make the dog come to you. If you do not enforce your commands, your dog learns that he does not always have to respond.

When Your Dog Challenges You

I must reiterate that if the dog growls, shows teeth, bites, or even threatens to bite, then he is challenging you on a core level. That dog wants to duel for your leadership position. Also, a dog that does not follow commands in other areas may not respect you as the top dog. Don't confuse this with a dog that simply has not had enough training and reinforcement. A dog will not come if he does not understand the command and has not had a substantial amount of training with positive reinforcement.

If you do experience relationship problems with your dog and he does not follow commands he knows, a restructuring of the owner-dog relationship is most likely in order. Review the five rules of rank and adhere to them more closely until the negative behavior ceases. Once your dog has structure in his life, being on the bed for a cuddle is a lovely reward that he has earned.

Put Your Foot Down

When I was a guest on *The Oprah Winfrey Show*, Oprah expressed concern over her chocolate Cocker Spaniel, Solomon. Solomon would not come when Oprah called him. I suggested she revoke Solomon's privileges, and since sleeping in the bed was one of his favorite things to do, I told her to kick him out. Oprah was a bit distressed by my suggestion and protested, saying he was fine in the bed; the bed was not the issue. The problem was when she called him. What frustrated her most was Solomon obeyed his trainer but not Oprah.

Oprah was right about one thing: the bed was not the issue. She and Solomon had a classic relationship problem. Solomon obeyed his trainer, not because he preferred her to Oprah, but because he thought the trainer ranked higher than he. Solomon did not view Oprah as top dog. He also had a reinforced behavioral history with the trainer that he did not have with Oprah. Solomon viewed the trainer as more of an authority figure. He knew he had to obey his trainer, whereas with Oprah he felt he was just as much in control as she. Also, because dogs are situational learners, they understand based on the context within which they are trained. Solomon listened to the person who spent the most time working with him: the trainer.

Revoking privileges is not permanent. It is a temporary action until you regain leverage in the relationship. When you put your foot down, your dog regards you more seriously. For many dogs, once they are more obedient, they may earn back their privileges. It may be fine to allow them back on the bed. Oprah's hesitation was a common reaction. Many people have a hard time understanding why their dogs do not obey them. Most of the time, it boils down to rank or inadequate reinforcement of good behavior. If you want your dog to listen to you, reinforce appropriate behavior, and make sure he understands you are the leader.

Rank in a Multi-Dog Household

Dogs are pack animals and need rank to maintain order. If a pack does not have a pecking order, chaos occurs. They look for a leader as did their canine ancestors. Dogs naturally establish a hierarchy through body language and the possession of coveted resources like food, toys, and resting areas. Most problems are caused in a multi-dog household when an owner tries to establish an egalitarian setting for the dogs. In a house with passive dogs, this may not create any struggle, but with confident or dominant breed types, aggression may be a big issue.

Supporting Pack Hierarchy

Most owners recognize natural dominance in one of their dogs. They feel bad when they see the submissive postures exhibited by the subordinate dog. They do not understand that this is the dog's way of maintaining peace. A dog that fails to go after a ball thrown in the presence of the dominant dog is simply deferring to the dominant dog. Deferment is correct behavior to maintain pack harmony. Trouble between dogs begins when people fail to recognize and support the dogs' natural hierarchy. To support the pack hierarchy, follow these guidelines:

- Lower the rank of all dogs in the household, and ensure your position as pack leader.
- Give all attention, praise, and privileges to the dominant dog.
- Feed and give treats to the dominant dog first.
- Back the dominant dog if there is a fight over anything—food, toys, bones, resting place.
- Defer to the dominant dog by greeting him first.
- Give attention and privileges to the subordinate dog when not in the presence of the dominant dog.

Owners sometimes feel the need to give the subordinate dog extra attention when the dominant dog takes charge. This could provoke an increase in aggression by the dominant dog, which may feel the subordinate dog is receiving undeserved privileges. It could also bolster the subordinate dog's confidence to a level where he may challenge the dominant dog, resulting in dog warfare. It's usually pretty obvious which dog is more dominant. The more dominant dog will do things like take possession of items such as toys, bones, and food; control passageways; and push the other dog out of the way when you pet him. Once you assess which dog is the more dominant dog, you should respect his position.

Sometimes dogs may be equally dominant and aggressive. These dogs may never be able to live together peacefully in the same household. Also, in rare instances, the dominant dog may be so overly aggressive that backing that dog could make him more of a bully. Ruby would naturally be much more dominant than Lotte. I have successfully gone out of my way to reinforce Lotte as top dog. Because Ruby is naturally a bully, subordinating her position helps me maintain my control over her. This was the best solution for my particular situation. If you are a dog owner in this situation, seek a professional trainer or behaviorist by calling the Association of Pet Dog Trainers at 800-PET DOGS (800-738-3647). They will refer you to a trainer and help you attain peace in your home.

Dogs Don't Do Democracy

Rank keeps harmony in a multi-dog house. If you do not support the hierarchy among your dogs, you risk losing a sane household. Such was the case with actress Ally Sheedy and her three dogs. Her dogs Jordy, a three-year-old neutered Lab mix, and Becky, a two-and-a-half-year-old spayed Australian Cattle Dog, were fighting with Buddy, her two-year-old neutered Cocker Spaniel.

Ally did not want to show favoritism, so she fed all the dogs at once, petted and gave attention to them simultaneously, and gave treats to all at the same time. She did, however, allow Buddy to sleep in bed with her. Ally was sending mixed messages to her dogs. She needed to recognize the number one dog and reinforce that dog's position.

Buddy was the most pushy and clearly the most dominant. We put Buddy first, Jordy second, and Becky last. By putting the food down in the hierarchical order, greeting the dogs in rank order, and ignoring the other dogs and purposely pushing them out of the way, we helped establish a pecking order among Ally's dogs. The more Ally supported Buddy's position, the faster the other dogs recognized his higher rank. The fighting among the dogs stopped, and Ally gained more control over her dogs.

Be a Benevolent Leader

Sometimes in canine packs, a higher-ranking dog or wolf will give a privilege normally reserved for the alpha to a lower-ranking pack member. This is not seen as a sign of weakness but rather as a benevolent gesture on the part of the dominant animal. Once we establish rank with our dogs, we may be able to allow our dogs privileges without losing our status. A kind act is not always viewed as a subordinance play by the dog. We can respond to our dogs' desires, knowing our place or role is intact.

A strong leader understands and capitalizes on what's essential to his dog. He uses it to his benefit. He plays games with his dog and gives him appropriate chew toys to satiate the dog's natural instinct. He is also aware that those same instincts can be problematic. Truly understanding the ramifications of what you've taken into your home is the best way to love and care for your dog. A lack of understanding is why we continually have problems with our dogs.

In Short

Dogs are either leaders or followers. Make sure *you* are the leader. Stand tall, use a voice with an authoritative tone, and make direct eye contact; the idea that you're the boss will be clear.

The five rules of rank are:

1. Leaders always have the choicest resting spot.
2. Leaders eat first, and they control the food.
3. Leaders control the giving of attention and affection.
4. Leaders make the rules and control the resources for play and games.
5. Leaders go first, controlling entrances and exits.

Initially, your dog should work for attention and praise. Nothing should be free. Always follow through on all commands given. Once rank is established, dogs can be treated to privileges.

In a multi-dog household, all the dogs must be comfortable with their place to live harmoniously. You must recognize and support the pack hierarchy.

A strong leader is also a benevolent leader who understands and empathizes with his dog.

6

Socialization

In the movie *Nell,* Jodie Foster plays a woman raised in the woods, isolated from the rest of the world, a wild child. The only people she has ever known are her twin sister and her mother, a stroke victim with a severe speech aphasia. She knows nothing of our modern conventions, fears the outdoors during the day, and has developed a language unique to her. The film questions if Nell, an adult woman, can function in society without ever having been exposed to it. Will she be able to adapt?

Nell is truly a study in human socialization. Humans are social creatures. No one will argue that point. From the day we are born, our socialization begins. The same thing can be said about dogs. Dogs, too, are social animals. They need to be taught social skills from day one just like us. If you want a well-adjusted dog that can live in your home and be part of your life, that can fit in with your family and be comfortable with your friends, it is imperative that you teach him from the earliest possible moment how to get along with other animals and humans of all shapes and sizes. Failure to do so is an invitation to trouble.

Forgive the pun, but in *Nell* the doctors had to teach an old dog new tricks—not an easy task, particularly when the subject fears all new stimuli. If you do not properly socialize your dog early on, you will be faced with a job similar to that of the doctors in the movie.

What Is Socialization?

Socialization is the familiarization or acclimation of animals, be they human or other, to persons, places, and things. In actuality, you do not socialize beings to locations or inanimate objects. You do, however, desensitize them. Your dog cannot be socialized with a trash can, but he can be made more comfortable around one. Socialization in its true form involves interaction between one being and another. However, for our purposes I will widen the term *socialization* to include the introduction to all novel stimuli, whether people, places, or things.

Why Socialize Your Dog?

Failure to socialize your dog can result in an aggressive, antisocial, or fearful adult dog. As an adult or adolescent, your dog may fear something he was exposed to as a young puppy. More likely, he may fear novel stimuli or something he was not exposed to as a puppy. The point is, a dog feels fear for vast and varied reasons. This is why proper socialization of your dog or puppy is so important.

Fear Is a Four-Letter Word

Fear overrides everything. It is one of the most difficult emotions to deal with because you cannot train through it. It is overwhelming and crippling to the dog. How can you train a dog to sit if you can't even get him to look at you because he is terrified by his surroundings or by you? You must allow the dog to become comfortable before you can actually begin to train him.

The Three F's

Like most animals, including humans, dogs have three reactions to fear: flight, freeze, or fight. These are the basic primal instincts to any given situation. The "flight" dog is the dog that avoids confrontation and runs from the situation as quickly as he can. The "freeze" dog will do just that—freeze, or possibly even faint—as a reaction to something overwhelmingly frightening. Finally, there is the "fight" dog. This dog fights back in fear when cornered or unable to get away from that which threatens him. Knowing this, we can make a direct connection between fear and aggression.

Fear-Induced Aggression

Fear aggression, or fear-induced aggression, is probably the most difficult type of fear to work through. It is also the most worrisome. Instead of shutting down or shying away from the source of the fear, a dog may instead snap, bite, or act aggressively toward it when unable to escape. These dogs can be unpredictable. Sometimes their aggressive behavior makes it virtually impossible to even reach the dog and begin working with the deeper issue of fear.

Fear-induced aggression is often a learned behavior. Once the dog growls, snaps, or snarls and sees the source of his fear move away, he learns that his aggressive reaction elicits the response he wants. Essentially the dog is saying, "Move away from me. You make me uncomfortable." The dog learns that aggression is an effective method to remove the scary source.

All Positive, No Negative

An owner should never respond to a dog's aggression with aggression. It worsens the situation. If you punish your dog for growling at a person, the dog only learns to censor his growling and snapping behavior. You may dangerously create a silent attacker by training the dog to suppress his warning signals. The next time something or someone scares him, he will not bother growling or snapping and instead may go straight for the bite.

By treating a fear-induced aggressive dog with aggression, you create more anxiety for the dog. The dog quickly learns to associate the scary person, place, or thing with a bash on the head. You compound the problem because the dog becomes more apprehensive about the source rather than more comfortable. To better the situation, use food and approximate distance. For example, if your dog is afraid of children, start with one child at a comfortable distance for your dog. Gradually move closer and closer to the child while continually rewarding your dog for every favorable response. This is all done with food, not punishment.

When to Socialize Your Dog

Studies show dogs need to be exposed to as many novel things, beings, and places as possible within the first few months of their lives. The first twelve weeks of a dog's life are truly the most formative. During this three-month period, the puppy learns about the world in which he lives. Failure to socialize your dog during this very crucial window of time could create future behavioral problems such as fear of anything or anyone the dog encounters or, even worse, biting. Don't forget: fear is very difficult to overcome and train through. It is imperative to introduce the puppy to as many different people, animals, objects, and places as possible during these first crucial weeks rather than isolating him in the microcosm of your home.

Start from Day One

Because they are pack animals, dogs are very social creatures. Like humans, they need to be taught social skills from the very beginning of their lives. This is one of the main reasons you need to find out everything about the dog's history before bringing him home. Ask the breeder how the dog has been cared for. Following the breeder questionnaire in Chapter 2 as a guideline should help give you some insight as to how much work has gone into socializing the dog. Although rescuing a dog is a wonderful and noble act, it can also be problematic because it is sometimes very difficult to find out how the dog was raised.

Delayed Socialization

People often assume that if a dog is afraid of someone, then he was probably abused or hurt by someone resembling that person. Of course this is a possibility, but it's more likely the dog fears that person because he was *not* exposed to that type, size, or color person at a young age. Delayed socialization, which is an attempt to socialize the dog after the twelve-week window of opportunity, can be a difficult and, in a few cases, futile effort.

Take Chloe, for example. Chloe is a Standard Poodle Jack Lemmon and his family purchased from a breeder. At the time, Chloe was only five months old, but that was already two months past the effective socialization period. Chloe was very shy and fearful, most likely because the breeder did not put in the effort to properly socialize her. She particularly feared the Lemmons' African-American assistant. This man was kind and gentle and wanted to learn to work with Chloe. He never treated her harshly, yet she feared him terribly.

We began by overcoming some typical behavior problems such as house soiling and chewing. Once those were under control, we began to acclimate Chloe to every member of the household by having each one walk her around the backyard on a leash while giving her treats. Initially, she was scared of everyone and everything, but eventually, she became comfortable with everyone. Everyone but the assistant. She remained terrified of him.

The Lemmons enrolled Chloe in puppy class where she was exposed to other people and dogs while experiencing a new environment. She shined in class. Jack Lemmon and Chloe took the third-place ribbon in puppy graduation. Yet regardless of how well Chloe did in class, she remained frightened of the assistant. Two years later, the Lemmons brought Chloe to me for boarding while they were out of town. They felt it was the best thing to do because she would otherwise be left alone with the assistant, whom she still feared to her very core, even though he walked her on a daily basis.

Despite all our efforts to socialize Chloe with this man, she remained frightened of him. He tried very hard to make friends with Chloe, but his tall, dark figure was too imposing for her. This shows how a mere five months is enough time to create lasting damage through improper socialization. The scars from a lack of socialization in the dog's formative period were so deep that she could not surpass them. True, she overcame several of her fears and much of her shyness; but sadly, she never felt comfortable with this friendly man.

An Ongoing Process

Just because you've exposed your dog to something as a puppy doesn't mean it ends there. Socialization never ends. I vividly remember training a Golden Retriever from the time she was three months to six months old. I took her for walks, exposing her to everything: children, trash cans, trucks, mailboxes. The dog managed herself exceedingly well. The owners never took a puppy class and stopped training when the dog was about six months old.

I saw the dog again approximately six months later. In that period, the owners never took her off their property. They brought her to me because they finally tried to walk her and found she would not leave the house. When at home, she was a very confident, almost dominant member of their family, which included another dog. Once off the property, the dog reeled back in terror at the sight of a trash can, a truck, a car, even other dogs. At one time, their Golden was socialized to everything she came to fear. The owners nullified her previous socialization by isolating her from the world and discontinuing her socialization.

Immunization vs. Socialization: The Great Debate

Puppies need a series of vaccinations so that it is safe for them to be around other dogs. The shots begin when the puppy is six weeks old and continue until he is four months old—the exact period during which the dog must be socialized. Many veterinarians recommend isolating your dog from other dogs until he has had his last shot. This means no walks, parks, beaches, puppy classes, or any other places where your puppy might encounter other dogs. Some vets go so far as to not allow the puppy into the owner's own backyard!

Obviously, this poses a dilemma. The need for socialization and the need for immunization occur simultaneously, yet many people believe those needs are mutually exclusive. Vaccinations are not 100 percent effective. Your dog can still catch a disease even after his last shot. And unless your dog lives in a bubble, he can be exposed to disease. Some diseases are airborne, or you can step in waste and unknowingly bring them into your home. This is why breeders make you take your shoes off when you visit a litter of puppies. The decision of whose advice to follow is yours, but I encourage you to find a compromise that combines both socialization and preventative health care.

Puppy Classes

I have never known a puppy that caught a disease while in a class. Of course, the possibility always exists. More and more veterinarians are beginning to understand the importance of early socialization, and they confidently refer dogs to a reputable puppy class. Still, there are veterinarians and breeders who argue to keep your puppy out of class. It is right to make the dog's health a priority. But socialization and behavior management are just as important as physical health. Your dog should be both physically *and* mentally sound. It's not fair to the dog to sacrifice one for the other.

Informed Risks

I often encounter people in this predicament: they want to enroll their puppy in a puppy class but fear the potential for disease. I have to agree, disease is a real problem. But if you use common sense and a bit of prudence, it is possible to successfully socialize your puppy and keep him healthy at the same time. For example:

- Enroll your puppy in a veterinarian-referred class. There you can be more confident your dog will be in the company of other dogs that are current in their immunization program.

- Don't visit dog parks or streets where stray animals may wander, but rather stick to your neighborhood, where you know the level of responsibility people take with their pets.

- Take your puppy to the home of a friend who you know has a healthy, immunized dog.

- Invite dogs you know are healthy and current in their shots over to your house.

- If possible, take the dog to malls, shopping centers, banks, or other places where there probably won't be any dogs. This way the puppy will become accustomed to people and places.

- Have a puppy party, where you invite friends into your home to meet your puppy.

If you are still reluctant about taking your puppy near other dogs before the immunization period is over, you should try to expose him to as many new people and places as possible. You can carry a small puppy or toy breed in a carrier such as a Sherpa bag, which is a mesh travel bag available in pet stores. This way you don't even have to set the puppy on the ground, and he can still look around and see people, dogs, and everything else in this great, big world.

This can be a difficult decision for many people. Weigh the options carefully. I've seen many dogs afraid of their own shadow because of poor socialization early in life. Take precautions, but ask yourself if being overcautious is worth potentially risking your dog's mental well-being. With Ruby, it was more important for me to take informed risks and chance infection than to have an aggressive and antisocial dog that I would never be able to take anywhere in the future.

Entering the Social Scene

There are several ways to help your puppy blossom into a social butterfly. Keep in mind that each dog is different, and that your dog may not handle a situation in the same manner as your best friend's dog. Be sensitive to your dog's personality as you introduce him to the world, but at the same time, do not baby him. Some of the following suggestions will help turn your dog into a debutante.

It's Party Time

A great way to increase your dog's social skills while maintaining your own is to host a puppy party. Invite your friends over to meet your new puppy, and make dinner for them as a draw. Dogs aren't the only ones who respond favorably to food! This is a good opportunity to see how your dog responds to new people while in a controlled environment. So stock the house with lots of puppy treats, and get ready for your puppy's first "coming out" party.

Pass the Puppy

The first game you play is *Pass the Puppy*. Hand out puppy treats to all your guests. Freeze-dried liver is the gold standard of dog treats. Whatever you choose, make sure it has an extra-special taste with which the puppy can associate the people he is about to befriend. Also, make sure you have not fed your puppy, and cut the treats into tiny pieces so that he does not get sick or full.

Next, have your friends hold the puppy, one by one. Each should individually feed your puppy a treat. Give the guests free rein to get as mushy as they want with the puppy. This shouldn't be a problem because people love puppies. They'll probably fight over who gets the puppy first. If the puppy freaks out and won't cooperate or take treats, back off. You could be doing more damage than good if you continue to pass him around. Always keep your puppy's personality and needs in mind. If he's a shy puppy, don't invite ten people over. Limit your guest list to one or two people.

If the puppy shows visible signs of distress from being around so many strangers, try introducing him to your friends more slowly. Rather than having a barrage of people passing the puppy around, have each person come to the puppy individually and spend some time with him. The puppy may not be coping if he will not take a treat, tucks his tail between his legs, or pulls away from the stranger. Your litmus test is whether or not the puppy takes the food from your guests. Always give the puppy the option to retreat and return when he is ready. Allowing the puppy a safe haven will prevent him from having an aggressive reaction to his source of fear. If the situation becomes too stressful for the puppy, stop the game. You could be taking steps backward by creating a dog with a fear of people.

Round-Robin Recall

If the first game goes off without a hitch, you can then move on to a game of *Round-Robin Recall*. Not only is this exercise great in terms of socialization, but it helps teach and reinforce the puppy's training skills by practicing the come command.

Again, pass out tiny treats to your guests. Gather everyone in a circle. One person should bend over and entice the puppy with a treat while calling to him, "Come, Henry, come." When the puppy responds, he should be given the treat and praised with either "good Henry" or "good come." The person should then turn away, totally disassociating himself from the puppy. Not even eye contact should be made with the puppy. Someone else in the circle starts the process all over again. Continue until everyone in the circle has a chance to call the puppy, reward and praise him when he responds, and then turn away from him.

Losing focus with the dog may seem counterproductive in the socialization process, when in fact, it is crucial. By disassociating yourself from the dog and losing focus, you encourage the puppy to move on to the next person. If you continue to stand there, engaging the puppy, he will most likely ignore the others in the circle. He knows he can get a treat from you. You're a sure thing. He's not so certain about the others, though. When you turn away from the puppy, he thinks the source of his treats is gone, and he will be more willing to bound over to the next person in hope of finding another yummy snack.

Going Out on the Town

The best way to expose your puppy to a variety of stimuli is to take him out on the town. Every day we pass a multitude of objects, places, and people to which we have been desensitized. A man with a beard walking down the street may not be novel to you or me, but to a puppy, he could be a brand new sight eliciting intrigue or fear. Out in the world, your dog can see, hear, and smell cars, buses, trucks, traffic lights, trash cans, fire hydrants, dogs, cats, mail carriers, people with strollers, people of different races or ethnic backgrounds, heavy people, thin people, people with hats, people with canes, people with packages or shopping bags, tall people, small people, old people, young people. The list is endless. These are all things your dog may not encounter on a daily basis, but sooner or later, your dog is bound to come in contact with someone or something that resembles one of these people or things.

Stepping Out

Just going for a walk is a fabulous way to expose your dog or puppy to a variety of elements. The most productive walk in terms of socialization and desensitization is one that takes you past an abundance of stimuli the dog might not see, smell, or hear walking down your own street. The more you take these walks, the better, since they increase your dog's confidence.

You can add another dimension to your walks that will pay off handsomely. When you walk by a "new" person, meaning someone you perceive as different from the people to whom your dog is accustomed, stop and ask that person to interact with your puppy. This usually is not a problem since most people enjoy meeting an adorable, little puppy. Explain that you want your dog to like everybody. Then ask the person to offer your puppy a treat, and pull out your delicious puppy snacks.

Many dogs are frightened by large men, young children, the disabled who use health-related equipment, elderly people, or anybody different from their owner, so try to approach those types of people while walking your dog or puppy. As before, try to get them to offer the dog a treat. That way, you can be certain you are pairing the look and smell of a new person with a positive association for the puppy.

Crashing Parties

When invited to someone's house for a social gathering, if appropriate, ask permission to bring along the newest addition to your family. The supplies you will need for a successful visit are: the crate for nap time; a leash, even though the puppy may be very young and you plan on carrying him around; treats, obviously; a bag to clean up after the puppy—you want to keep the yard clean so you're both invited back; a bottle of odor neutralizer, in case your excited puppy has an indoor accident; and a chew toy, so the puppy keeps his mouth off other guests and furnishings.

The first thing you do when you arrive is immediately take the puppy outside to an appropriate area isolated from the street. Put the puppy down and give him the potty command. Once he goes, you've established a potty area. If he does not go, then carry him or at least leash him and take him out in another ten or fifteen minutes. When you make the introductions, ask the people if they mind feeding your puppy a treat, and give them one of the snacks you brought along. Hold the puppy while the guest offers him the treat.

If your puppy is reluctant, cajole him with your voice and marvel over the presence of the treat. If the puppy continues to resist, back off. The new environment and new people might be too stressful for the puppy to handle all at once. If this happens, find a quiet, isolated place and take the puppy there. Once the two of you are in your own space, feed him treats and try to relax him. When he calms, take the puppy back to the party, but ask the guests not to focus on him. If the puppy seems to adjust, you might try to have the guests feed him treats again. By doing this, the guests let your puppy know they are friendly.

Cruisin'

One way to get your puppy into the world is in the car. Your dog can hang out in the car and check out all the sights as you drive through town and run your errands. Make sure the dog is confined, either harnessed to seat belts, crated, or gated in back. Always try to take the dog with you as long as you can leave him safely in the car and preferably in his crate. Remember, the inside of a car is usually many degrees warmer than the outside, especially in the summer. Make sure the temperature is not too hot before you leave your dog in the car, and keep your time away short. When beginning, this is most safely done with a partner who can stay in the car with the dog to monitor both the temperature and the dog's response to your leaving.

Many puppies get car sick, and their owners make the common mistake of avoiding taking the puppy in the car altogether. They assume it is too unpleasant for the puppy, and cleaning up is not exactly a joy for them either. They only take the puppy for a ride when they absolutely have to, which usually means a trip to the vet, groomer, or other destination the dog may not enjoy. Unbeknownst to them, they condition the dog to associate the car ride with an unpleasant event.

Start out by taking the dog on quick, nearby errands. The dog can probably handle that short amount of time, and you can often see the dog from wherever you are. If you are going to the movies or the mall for a couple of hours, don't even think of leaving the dog in the car. If you're worried about the dog getting sick in the car, take him on an empty stomach: no food or water for at least four hours before you go. It's also a good idea to put towels down or crate him; if your dog does get sick, it's easier to clean.

The first time I took Lotte to the market, I parked the car in a space close to the store where I could watch her from the window. I went in for a minute, and she started crying. I stayed in the store, and she gradually got used to the car. Now she does very well in the car. Not only does she feel protected in it, but I feel safe knowing she's protecting my car.

Love Thy Stranger

All these exercises prepare the dog to be in the company of strangers. The puppy learns people are not threatening and are to be accepted readily. Some people feel uncomfortable with this. They argue that they want their dog to be wary of strangers and to bark when a stranger comes to their door. That makes sense. It is the reason many people get dogs. One of the reasons I got Ruby was to protect my home. However, you have to ask yourself if you want the liability risk of owning a dog that attacks your friends as well as strangers who knock at your door.

Besides, once you allow a person into your home, your dog sees you have accepted that person. If the person changes his tone or attitude and becomes menacing or makes aggressive overtures, your dog will sense that, too, and respond accordingly. Socializing your dog will not affect his ability to be a good guard dog. In fact, it strengthens the dog's ability by making him a more discriminating sentry.

If you don't socialize your dog, you run the risk of having an animal that may cower in fear whenever anyone comes to the door. Or worse, he might be wary of all people and will lunge, bite, or bark indiscriminately. You could create a menacing dog, causing a multitude of problems for yourself that may even result in your having to put the dog to sleep.

Food Is the Key

The success of these exercises hinges on the use of dog treats. Some people do not feel comfortable allowing others to feed their dog and may even boast about how their dog will not accept food from strangers. They feel a person might feed their dog something bad or poisonous. Sure, it is possible, but realistically, the chances of this happening are slim. By reprimanding your dog for approaching or taking food from a stranger, you are only teaching him to fear other people.

Remember, the three primal reactions an animal has to any fearful situation are flight, freeze, or fight. Please keep this in mind if you have any hesitation about socializing your puppy.

We Shall Overcome: Coping with Fear

Before you can take your puppy for a ride, you have to be able to get him in the car. This may not be such an easy task if the dog is afraid of cars. The same stands true for any behavior where fear is involved. If your dog is afraid of garbage cans, how can you take him for a walk on trash day, when the street is lined with them? Three methods you can try to help your dog overcome his fears are flooding, systematic desensitization and counterconditioning, and countercommanding.

Flooding

Flooding is exactly what it sounds like. You flood the dog with whatever stimulus frightens him. When the dog is afraid of the car, you flood him with exposure to the car. You drive and drive and drive. Similarly, if the puppy is afraid of people, take him someplace where he is surrounded by many people. Be warned that flooding can be counterproductive. You could be doing more damage by increasing the dog's level of fear. This method is for certain dogs and certain situations.

Sometimes a dog in a flooded situation is inhibited by all that's going on around him. He may not react. However, the same dog may have a problem in a one-on-one situation. For example, an aggressive dog in a puppy class may behave, whereas if he encounters another dog on the street, he may react very aggressively. His reaction will be greatly related to the stress level created by the situation. This method is most effective in dogs with low levels of fear and a greater ability to bounce back.

Counterconditioning and Systematic Desensitization

Counterconditioning is generally used in conjunction with systematic desensitization. Counterconditioning and systematic desensitization are slow processes where you gradually acclimate the dog to the source of the fear and change a fearful or negative association to a positive one. Let's look again at the movie *Nell*. Nell was afraid to leave her house in the daytime. Her doctor slowly eased her out of this fear using food. First he gave her some popcorn, a food she had never before tasted. She loved it.

She wanted more, but he moved off the porch. Even though she was terrified to walk outside, her desire for the popcorn was so strong, it overpowered her fear. He continued to move farther away, taunting her with the food, and she continued to follow until she found herself standing safely in the bright sunlight.

Similarly, you can help your puppy overcome his fear. The use of treats is extremely powerful. Dogs respond very well to food. Let's say your puppy is afraid of cars. You can't even get him near a car, let alone in one. Take him to a distance where he can see a car but still feel comfortable. Give him a treat. Move a little closer to the car, and give him another treat. Keep moving closer and closer to the car, rewarding him for each bit of progress. Slowly, your puppy will feel more comfortable in the presence of cars.

With a dog that does not like to be inside a car, you might want to try feeding him, giving him treats, or handing him toys in the car. You might even want to crate the dog in the car if the crate fits. Take him on short, fun trips to places like the park. If the dog is really scared, start out slower by letting him sit in the car. When he's fine with that, turn the car on, but don't go anywhere. Then try rolling out of the driveway. The next step would be driving to the end of the block. Get the picture?

I'm using a car as an example, but this process applies to pretty much anything or anyone a dog may fear. If at any time the dog refuses to take a treat, then slow down. Take him back to a point where he is comfortable, and try again to move forward. Eventually, you will break through the barrier. Treats work well because they help substitute a positive association with the source of fear. Try to think of other ways to create positive associations for the dog.

What's That You Say?

There are many facets of daily life we take for granted that our dogs do not. Sound is a prime example. A dog's hearing is far more sensitive than ours. Your dog may not be as accepting of the wide range of sounds you know as commonplace.

When the puppy is young, you might want to keep the television on so he gets used to a variety of noises. Immerse him in sound. MTV works well because it has a lot of loud noises. If the cacophony on your TV scares your dog when you start, then ease up. Give the dog a treat as you turn on the TV, or better yet, feed him while the TV is on. Start at a lower volume, and then work your way up as the dog becomes more comfortable with the sound.

Some loud noises are more likely to send a chill up your dog's back than others. Washing machines, blenders, hair dryers, vacuum cleaners, and lawn mowers are at the top of the list. If you turn on the vacuum cleaner and your dog runs for cover, you better believe you have some serious desensitizing ahead of you.

Start by turning on the vacuum in a room far from the dog. Feed the dog while the vacuum runs. If the dog eats without a problem, then repeat the process the following day with the vacuum a few feet closer to the dog. Repeat this daily until you can bring the vacuum within close range of the dog. If the dog is too stressed to eat while the vacuum is on, you have it too close to him.

The success of this exercise depends on how skittish your dog is with the noise. It's also not limited to vacuum cleaners. You can use this exercise for desensitizing your dog to any loud noises that occur daily in your home.

Countercommanding

Countercommanding is generally used in conjunction with systematic desensitization. With countercommanding, you distract the dog by giving him an incompatible behavior. He cannot jump if he's sitting. Let's say your puppy is afraid of garbage trucks. Every time a garbage truck passes by, he panics. What do you do? Take his mind off the garbage truck by giving him something else to do. When that big truck comes rambling down your street, tell your puppy to sit. Once he sits, give him a treat. Remember to always start at a comfortable distance for your dog. If you continue this process, eventually you will condition your puppy to have a new association to the source of fear, sitting calmly rather than panicking, whenever he sees a garbage truck. He will no longer think about the truck and instead will concentrate on sitting and receiving his treat.

Counter the dog's fearful reaction with a new behavior. Be careful, though. When you countercondition a behavior, you must never elicit a fearful reaction. Don't simply give your dog a treat or try to soothe him with your voice. This could actually backfire because the dog might think you are rewarding him for his fearful behavior. By directing him to perform a physical behavior such as the sit command, the dog associates the reward with "good sit."

Never Too Soon

It is easier to work through fear with a puppy that is twelve weeks or younger. It is much more difficult when the dog is any older. Remember Jack Lemmon's dog, Chloe? She was only five months old, and irreparable damage had already been done. Delayed socialization can work, but it requires a tremendous amount of time and patience. Yet another reminder: the window of opportunity to most easily socialize your puppy is restricted to the first twelve weeks of his life.

Nature vs. Nurture

A puppy's mental and physical makeup is always a combination of genetics and environmental factors. Sometimes you can socialize your puppy to no end. Even if you start the process when the puppy is at a very young age, you may still end up with an antisocial, aggressive dog. Why? Because it was written in his DNA. It could also be that he learned it from his mother. It's the age-old battle between how much genetic information and how much environmental influences affect our lives in terms of our behavior.

Problems and Misconceptions

The following are some common misconceptions about training:

- **I don't want to train a puppy, so I'll get a slightly older dog.** One of my clients wanted to get a Dachshund, but she did not want to train a new puppy. She wanted a young dog that was about a year old. I advised her to get a puppy or let me help her choose an adult dog because some breeders do not properly socialize their dogs. Breeders who have a lot of dogs may not afford each the time he needs. A good trainer can help you weed through the dogs and try to find the one that best meets your needs and lifestyle. Getting a dog anytime after his first twelve weeks raises the potential for socialization problems.

  My client, against my better judgment, purchased a year-old Dachshund from a breeder. Sure enough, when she and her husband called to schedule a training session, her husband told me the dog was scared of everything. I reiterated that with older dogs, even year-old dogs, there exists a strong possibility for problems. I have seen many dogs like theirs, with the odds stacked against them, turn around. This shy, fearful Dachshund suffered from what is called kennelosis, or what I've come to call the "Nell syndrome."

  Because Jodie Foster's character in *Nell* was isolated from the outside world, it terrified her. She was comfortable only in her own little world and ways of living. Similarly, dogs raised in kennels have not been exposed to the proper outside stimuli such as people, places, objects, and other animals. When they do finally encounter novel stimuli, they become fearful and express that fear in any number of ways.

  Ironically, this client did not want to do all the work training a new puppy entails; yet she ended up with a dog requiring more effort and training than a young puppy. There is no question that a properly raised, year-old dog would definitely be easier to manage than a young puppy. I cannot stress enough how important it is to learn all you can about your dog's history before you bring him home.

Of course there are success stories. I had an eight-month-old Jack Russell Terrier in one of my puppy classes. When she first arrived she was terrified of everything and wanted to leave. She scratched at and tried to crawl up the owner's leg in an attempt to be rescued. After only five sessions using desensitization, flooding, counterconditioning, and teaching her commands the owner practiced at home, the dog became comfortable enough in class to interact with some of the other dogs. After we employed all the tricks of the trade, she took first place at class graduation.

Often older dogs from a breeder are dog-friendly because they've been socialized with other dogs in the kennel. However, these other dogs are often of the same breed. The dog only likes other Dobermans or Wheaten Terriers or Poodles. If you bring the dog around a different breed, he screams bloody murder. Similarly, a new puppy that is not exposed to a variety of stimuli shudders in fear when he does finally encounter new places, people, dogs, or objects.

- **My new dog gets along fine with my other dogs, so he should get along well with all dogs he encounters.**

Successfully socializing your dog with a resident dog does not automatically guarantee that he will be socialized with other dogs. Sure, your dogs may get along fine, but how do they react to other dogs? Your dogs have bonded. They have a very personal relationship, but that does not mean they are comfortable with other dogs.

One of my clients brought a rescue dog into her home where she had three whippets. The rescue dog was a big-eared shepherd mix that flopped around with her three very refined and pristine Whippets. It was quite a sight. Even though the dog looked like a misfit among the group, he was totally content with his pack of Whippets. The second he saw another dog, he screamed, tucked his tail, and ran for cover.

- **It's better to get littermates.**

  Again, in the movie *Nell* they talk about twin-speak, which is an interpersonal communication between twins. Twins have a very intimate way of relating to each other that others cannot understand. Many people choose to get two dogs from the same litter, thinking this will create an overall better situation for all involved, dogs and humans.

  In some cases, it is better. If you are someone who works twelve-hour days and the dog will be left alone for extended periods of time, then of course it is better for the dog to have a canine companion. No question. Perhaps a better solution is not to get a dog in the first place. Be aware that with littermates, you are going to lose in the relationship. The dogs will bond much more closely with each other than they will with you. They will communicate in a way similar to twin-speak.

  Tony Krantz, a television executive, came to me with his Labradors, Oscar and Miles. We were training them on a basketball court, and the two dogs, only twenty feet apart, longed for each other so much they were nearly impossible to work with. They couldn't stand being separated a short distance.

  You may be thinking, "Well that's fine with me. In fact, it's kind of cute how they want to be together all the time." Besides, you're never going to have them apart. Right? Think about this. What if one has to go to the vet? What if one becomes sick and dies? Suddenly, the other dog is left alone and, more times than not, cannot cope without his sibling. It is not cute to do this to your dogs. It is cruel. By allowing your dog to overbond with his sibling, you have not made these animals separate entities but instead made them so dependent on each other that they have trouble surviving apart.

  If you do get littermates or dogs of a close age, initially it is best to separate them as much as possible, at least several hours a day. Ideally, they should sleep in separate crates in separate rooms and should have limited access to each other. This means twice as much work for you. Walks, training, outings, and exercise should all be done separately. You want the dogs to spend individual time bonding with you. As they develop their own personalities and relationships with you, then they can gradually spend more time together. Eventually, they will even be allowed to sleep together.

A Lively Outlook

I hope you can see a pattern developing in what I'm saying. Isolation and overbonding are two of the biggest problems people create for themselves when training their dogs. Isolation means keeping your dog separated from any outside sources or influences. By doing this, you create a tendency for the dog to bond only with you, your family, or other dogs in the house. As a result, you open the door to fear, allowing it to control your dog.

Socialization is a lifelong process. Your dog will continue to need your guidance as he adjusts to the many changes in his life. You may not see immediate results, but stick with it. The more you work with your dog, the more secure he will feel in all situations. By helping your dog build his self-confidence, you will both benefit throughout your lives.

In Short

It is imperative to teach a dog from the earliest moment the social skills he will need to get along with humans and animals in all types of environments. If you do not, the result may very well be a fearful, possibly aggressive dog.

A dog in the grip of fear cannot be trained. The only thoughts on his mind are flight or fight.

Fear cannot be punished out of dogs. All that results is a dog that stops displaying warning signs and instead bites without warning. To combat the fear response, turn to counterconditioning and desensitization.

To prevent fearful adult dogs, socialize a young puppy early and often. If you are committed to early socialization, you may find yourself at odds with your veterinarian over when it is immunologically safe for your puppy to meet the world. Discuss the health risks and weigh them against the benefits of early socialization. Come up with a compromise with which you feel comfortable.

Here are some possible socialization compromises:

- Enroll in a veterinarian-referred puppy class.
- Visit a friend who has a healthy, fully vaccinated older dog.
- Invite healthy neighbor dogs and their owners to your home.
- If possible, take your puppy to shopping centers, banks, or other places of business where it is unlikely there will be other dogs.
- Throw a puppy party.

WARNING: Avoid visits to dog parks, neighborhood parks, or other areas of high canine traffic.

Through games, parties, car rides, and other types of outings, your new puppy will learn to feel comfortable with his place in your world.

Keep in mind, socialization is a lifelong process.

Part III
Training

Sit, Stay, and All That Good Stuff

7

Training Theory

The world of dog training somewhat parallels the world of child rearing. For years, children were raised with the Victorian mindset of "spare the rod, spoil the child." Then, in the 1960s, philosophies on raising children moved toward "spoil the rod, spare the child." It was thought that we should not reprimand children. Instead, we should let them be themselves. Child-rearing theories have swung from one extreme to the other and are now searching for a middle ground.

Parents are beginning to move back toward the use of discipline. A child who throws a temper tantrum these days is more likely to be reprimanded or given a time-out. This is not to say they are beaten or abused as punishment. It means parents no longer ignore and tolerate *all* bad behavior, but instead recognize it and respond to it. There is a balance between love and discipline. Both reward and punishment are done swiftly, purposefully, and with love.

Evolution of Training

The dog world experienced an evolution similar to the one in child rearing. Formal dog training originated in the military. Armies trained their dogs with harsh and rigid techniques. As dog training became more mainstream and moved into homes, it was fashioned after militaristic training. If dog training was harsh, it was because of its origin, not because that type of severity is required to train a dog.

Behavioral science has revolutionized dog training. For the first time, we delve not only into how dogs think but also into how dogs learn. We finally have substance to draw from when shaping dog training. We no longer have to think that strict domination of the dog is the only way to get compliance from him. Behaviorists study how dogs interrelate and respond to others and their environment. Research has found that dogs' actions directly relate to what is important to them. Pack and rank are at the top of the list. The second part, animal-learning theory, looks at how dogs learn, which is primarily via responses conditioned through positive and negative reinforcement.

The combination of understanding dog behavior and animal-learning theory headed a new movement in training. Theoretically, we should train motivationally, rewarding proper behavior and, in many cases, ignoring bad behavior. Some trainers have interpreted this to mean you should never do anything negative to the dog. Not even utter a harsh word. This was an overreaction to the old-style training. It's imperative to find a happy medium between the two in the same way people raising children today seek to find a balance between two very different methodologies.

Basics for Survival

As surrogate parents, it is our responsibility to teach our dogs the basics for survival. This is why we need to train them. When training, use positive reinforcement by rewarding good behavior and add a side of punishment only when deemed necessary. You should primarily focus on rewarding good behavior, but on occasion you may bring in an appropriate reprimand. With a dog that understands the command but chooses not to obey it, a reprimand, if applied properly, is extremely powerful. Many people have a hard time punishing their dogs. The fact is dogs reprimand each other.

From the time a puppy is born, he is disciplined by his mother. The mother dog subjects her puppies to rules of behavior: if the puppy comes to her for a feeding and she's not ready, she'll snap at him and send him on his way; if he bites at her while nursing, she'll nip at him and grab his muzzle; if he gets too rambunctious with his littermates, she'll growl and put the dog in his place. Life for all of us is directly tied to learning what we can and cannot do. Survival in the wild is full of boundaries and discipline for young animals. Failure to learn results in death. That is what life is about. The same is true for dogs in our homes. Failure to become housebroken or to stop running into the street may result in the dog being given away to a shelter or getting hit by a car.

Old School

Because obedience training originated in the military, it relied on aggressive methods. Also, many trainers were not yet familiar with animal behavior or learning theory. The information was available, but it had not trickled down to dog trainers. Coercion and brute force were frequently used to teach commands and to correct behaviors.

A trainer teaching a dog to sit physically pushed down the dog's rear while pulling up on a choke chain. If when walking, the dog ignored a command to sit and heel, the trainer jerked on the leash until the dog complied. The dog learned by trial and error what was the correct response to avoid the pain.

Another behavioral problem, digging, had its own unique form of discipline. The trainer dug a hole or used the existing hole, filled it with water, and submerged the dog's head in the water. Compassionate, don't you think?

These techniques were used by many highly esteemed trainers. Training was so harsh, dogs could not begin their training regimen until they were six months old. They could not withstand the brutal punishments prior to that age. The fallacy of not training a dog until he is six months old was born of this. Delaying training had nothing to do with the dog's learning abilities and everything to do with the training methodology. We now know waiting until six months of age wastes prime learning opportunities.

The Broken Spirit

The primary goal of aversive training was to coerce the dog into obeying a command. This was euphemistically termed "love, praise, and affection" training and is still widely used today. Theoretically, the dog is supposed to work for praise such as a pat on the head and a few kind words. In actuality, the dog works to avoid a harsh correction and to achieve the cessation of pain. Of course he's happy to receive a pat on the head when the alternative is strangulation. Dogs learned fast that they had better do as they are told or else. The result: a trained dog with no spirit. This is precisely the origin of the equation TRAINING = BROKEN SPIRIT. Training should mean nothing of the sort. When done properly, training means communication, allegiance, and a strong bond with your dog.

Sometimes, when fear and aggression are involved, the use of force results in a problem more serious than the one with which you started. For example, if you jerk on the leash of a timid dog, you will further frighten him, causing him to recoil, cower, and possibly shut down altogether. An aggressive dog may retaliate by biting you. Either way you add fuel to the fire. You cannot train through fear. The worst thing you can do is attempt to instill fear in an already timid or aggressive dog.

New School

Motivational training employs the use of rewards to reinforce good behavior and tells us to disregard all bad behavior. A reward is something that increases the frequency of an action. For some dogs, a reward may mean a toy. For others it means food. In fact, for most dogs, food is the best reward. Supposedly somewhere out there exists a dog that will work and be happy when you pat him on the head and say "good dog." I have yet to meet this dog. A couple of nice words and a pat on the head are not enough to make most dogs stop chasing squirrels and come when you call them.

You need to find whatever rivals competing motivation. Competing motivation is anything that interests the dog other than you. It's a scent on the ground, another person, another dog, a squirrel. You can pretty much bet something else will appeal to your dog much more than you. The truth is, we all work for something. Whether it be money, food, sex, or status, some goal motivates us. It's the same with dogs. A dog needs to have something that makes him want to work. Almost all the time, that something is food.

Motivational training in its pure form has its roots in marine mammal training. These animals are far too large to be trained with aversive methods. You can't exactly coerce a killer whale into jumping out of the water. You can lure it and reward it with food when it obeys a command. It made sense to try this with dogs. Many people resist using food in training, but once they see the results, they understand its effectiveness.

As wonderful and enlightening as motivational training may be, in its idealistic, pure form, it is not problem-free. Ignoring ALL bad behavior requires a substantial amount of time and patience. It's not always easy to only reward good behavior and ignore bad behavior. Yes, it is feasible, but it can require an unbelievable amount of time and patience. It is usually not as problematic with marine mammals because staffs are hired to train those animals in that manner all day long. Also, when training sessions end, the animals are contained in their practice tanks, not left loose in their homes. That is just not practical for people and their pets.

My School

Absolutely, hands down, no question, use food in training. It is fast, fun, user-friendly, and foolproof. Everyone can do it, whereas not everyone can necessarily use a choke chain (nor should they). It's not very hard to hand a dog a liver treat or a piece of hot dog. Using food in training is a win-win situation. If your timing is off, the worst thing that happens is the dog gets a hot dog at the wrong time. The ramifications of improperly timing a treat as opposed to a harsh correction are much less serious.

Training is comprised of two things: (1) Does the dog know the behavior? And (2) does the dog respect you enough to do the behavior no matter what?

Does the dog know how to sit? You do umpteen repetitions and reward your dog appropriately. The dog sits in the park with squirrels. He sits with other dogs around him or children playing nearby. He sits in multiple situations. He knows how to sit. Will the dog sit for you whenever you want him to, just because you said so? That's where relationship comes into the picture.

You may not always have a treat handy. There comes a point where you must have enough of a relationship with your dog that you are undoubtedly his pack leader. Even if the dog does not want to listen to you at that moment, you must have enough control over the animal to get him to do what you want. Sometimes the dog needs a correction to establish that control. This is where I differ from the purist motivational method. I firmly believe healthy, effective dog training combines a barrage of positive reinforcement with a little bit of appropriately applied correction.

Ultimately, dog training boils down to a formula:

REQUEST + ACTION = CONSEQUENCE

It works both ways. You ask your dog to sit. He sits. He gets a reward. The reward is the consequence. Conversely, you ask your dog to sit. He does not sit. He gets a leash pop, thinks "ouch," and sits. Request, action, and consequence. There are a couple of other elements that factor into the formula: dogs live in the moment, and they do not generalize.

Living in the Here and Now

It is imperative that corrections as well as rewards happen during the behavior. Dogs live in the moment. They are creatures that do not think about what happened in the past or what will happen in the future. This means there should be *no punishment after the fact*. If you come home and find the dog has eaten a corner of your rug, there's nothing you can do about it other than ask yourself what you could have done to prevent this. Is the dog getting enough exercise? Do you give him chew toys and play with him? Is the dog bored? Are you simply giving him too much freedom? You are hardly exempt from blame, so don't put any of it on the dog.

No Punishment After the Fact

If you punish the dog anytime after the behavior has happened, the dog thinks he is being punished for what he is doing at that particular moment. Unless you catch the dog chewing on the rug, you are too late. If the dog is sleeping when you come home and find your rug chewed and you punish him then, he thinks he is being punished for sleeping. Not only will he not link the punishment to the earlier behavior, but he will also come to fear your homecoming.

Now the dog never knows who is coming through the door. Is it nice owner or psycho owner? This can lead to more anxiety and can exacerbate the situation. What does he do to alleviate the anxiety? He continues the behavior because he is stressed out about his impending doom and chewing relieves tension.

Your dog understands you are mad at him, but he does not quite comprehend why. Many people say, "Look at him. He knows he was bad," as the dog cowers away with his ears back and his tail tucked. With that submissive body language, the dog looks like he is feeling guilty. He is not. Dogs do not feel guilt. The dog is not thinking, "If only I hadn't done that." He is frightened and responding to your anger. In pack language, once an animal shows submission, reprimands cease. We perform the ultimate act of cruelty by continuing to reprimand the dog after he has waved the white flag.

The only way a dog understands and puts together what you are trying to say to him is if you disrupt a current behavior. Only then does he realize you are protesting what he is doing at that moment or, conversely, you are praising what he is doing at that moment. You give the dog the reward and say "good sit" the second his rear hits the ground following a "sit" command. If the dog is standing by the time you get the treat to his mouth, you are rewarding the dog for standing, not sitting. Rewarding good behavior as well as correcting inappropriate behavior should happen at the moment of the behavior. You have to be very, very specific and think about what the dog is doing at that exact moment so you can either reward or reprimand at that time.

The Pencil Test

For those of you who still believe the dog knows what he did was wrong and is feeling guilty about it, let me illustrate a simple little test.

Take any object—a box of pencils works fine. Place the box on the floor near your dog. Begin to reprimand your dog. Use the phrase of your choice—"off," "no," "bad dog." Once the dog tucks his tail, puts his ears back, or shows any signs of submission, remove the box of pencils.

The next day, place the box of pencils on the floor. If your correction was harsh enough, your dog will take one look at you, look at the pencils, and become submissive. According to your guilt theory, he must be feeling extremely guilty about whatever he's done to that box of pencils. Why else would he take on that body posture?

P.S. Don't try this. It would be mean and confusing. Please just take my word that this is the truth.

Instructive Reprimands

When you do catch your dog in the middle of a crime, give an instructional reprimand. Be specific. Don't just yell, "NO!" It is much more fair and effective to use an instructive reprimand. An instructive reprimand not only stops the behavior but also tells the dog what he should do. "No" tends to become too generalized.

If you catch the dog eliminating in the house, don't say "no" or "bad dog." Tell the dog "outside." If you say "no," the dog will stop the behavior, but he will still have to pee. By telling the dog "outside," you're telling the dog what he should do. It's much more instructional and much more fair. If the dog is barking, you don't say "no," you say "quiet." This tells the dog what he should do. He should become quiet. If the dog is jumping on a person, digging in the garden, or chewing on a shoe, you say "off," and the dog should remove himself from that person, place, or thing.

Dogs have been trained with "no" forever, and yes, this works, but you might as well take advantage of the fact that dogs can learn over two hundred commands. The more vocabulary you teach your dog, the greater the communication and the more enriched the relationship. The number of instructive reprimands you can use is limited only by your imagination. This is more than a good way to teach your dog what is appropriate; it's a fantastic way to increase communication with your dog.

Dogs Don't Generalize

Again, dogs do not generalize. They are situational learners. The context and location in which a dog learns something is how and where he will best exhibit that behavior. One of the most common complaints I hear in dog class when a dog will not comply is, "He does it at home." Obviously. That is where the dog is trained the most and where there is a much lower level of distraction. Until you expose the dog to many different places, distractions, and situations, including changes in the time of day, he will not generalize the behavior.

How do you overcome this? Really make an effort to train your dog in a variety of surroundings at different times of the day. As your dog becomes more cosmopolitan, he will be more familiar with a variety of situations. This is why dog classes are so great. The dog learns to obey you no matter what the distraction.

Take housebreaking, for example. Your dog, Lulu, is perfectly trained at home. She's never once had a potty accident in the house. You take her over to your mother's house one afternoon, and the first thing she does when you get there is pee in the living room. Embarrassed, you apologize to your mother and defend Lulu by letting Mom know she's never done that at home. Of course she hasn't. Lulu was trained at home. She has not generalized housebreaking to mean all indoor spaces. She knows not to go in your particular indoor space, but Mom's place is new to her.

Put the Burden on Yourself

It's rather brash of you to assume that because an animal is trained in your home, he will be the same sweet angel everywhere else. The proper way to handle bringing your dog into a new environment is to immediately familiarize him with his surroundings. Start by taking him outside to an approved potty spot, and tell him to go potty. Then watch him very closely as he scopes out the new scenery. Put the burden on yourself. We really do expect so much from our dogs. We place extremely high expectations on them, and when they fail, we are disappointed. Please realize, they are dogs, not little humans. Do not penalize them for not meeting your own unrealistic expectations.

Specific Posture

Do you think your dog knows how to sit? Here's a test for you. Try lying on the ground and telling your dog to sit. If the dog has never done a sit with you lying down, he will not have any idea what you want him to do. You will really confuse him. Dogs form a mental picture of what you look like when you give a command. When you alter that image, the dog has a really hard time connecting your new body posture with the command. A well-trained dog that is somewhat worldly will probably understand what you want with an easy command like sit. If you try a harder command like down, the dog will most likely have no comprehension of what you want. The dog will probably feel like he was suddenly dropped into a freaky episode of *The Twilight Zone*. That is how specific they are.

Will Work for Food

Positive reinforcement using food is the most powerful training technique. It is in a dog's nature to work for his food. By using food in training, we merely harness this natural desire. When you reward your dog with treats, break the treat into many small pieces. You can reward more repetitions, and your dog won't get sick from overeating.

Many times when I see new clients, they immediately and joyfully inform me that they taught their dog to sit. When I ask how, they sheepishly tell me "Well, I gave her a biscuit." There is absolutely no reason to feel ashamed about rewarding a dog with food. It is a kind way to train as opposed to hitting the dog or jerking him with a choke chain.

Create a Secondary Reinforcer

It is important that you create a secondary reinforcer—the association of one action with another. You want the dog to associate a word with the reward. For example, say "good" before you give the food treat so that later "good" alone has a very powerful meaning. A secondary reinforcer, such as the word "good," acts as a bridge between a favorable response and a reward. After enough conditioning, the word itself becomes a learned reinforcer or reward. When I was in the competition obedience ring with Lotte, where food is not allowed, I would say "good" and could see Lotte picturing turkey hot dogs in her mind.

Lure-Reward Training

The lure-reward technique is simple. Initially, you get the dog into position by physically luring him with a treat in a way that when the dog's nose follows it, his body naturally assumes the position you seek. Then you give him the treat as a reward. Eventually, you hide the treat from the dog so it is given only as a reward. Your hand becomes the lure, and after enough conditioned trials, the dog responds to a verbal command without the crutch of the hand signal. The treat goes from a lure to a reward, hence the name lure-reward training. Follow these steps to lure-reward training:

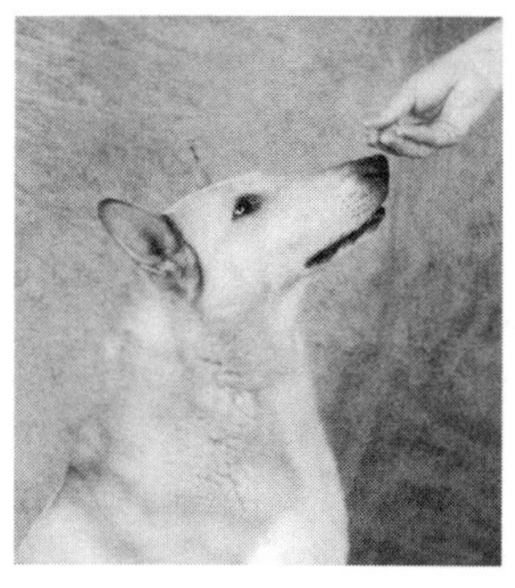

Lure-reward training with treat visible

1. Use the treat as a visible lure. The dog follows the treat into position and is rewarded with it.
2. Hide the treat and use your hand as the lure. Reward the dog with the hidden treat for compliance.
3. Phase out the hand lure and use the verbal cue only. Continue to reward the dog with the hidden treat.

4. Vary the number of treat rewards, reducing them to occasional rewards.

Timing the Treats

When first using food in training, reward every correct behavior. Soon you will begin to vary the times you give a treat by only rewarding the best efforts, such as initial or quick responses. By randomly rewarding the dog for his best efforts, you maximize excitement for him, and you shape the behavior exactly the way you want it. The dog learns that faster is better. Don't just settle for a dog that comes, but one that comes quickly.

The Jackpot

If you got a dime every time you played the slots, you would be bored. It would be too predictable. Never knowing what you're going to get keeps you hooked and coming back for more. You may not remember the hours you spend dumping quarters into the machine, but you do remember the one time you hit the jackpot. The same principle applies to treat training your dog.

For a breakthrough or a great effort, give the dog lots of treats. Shower him with treats like he struck gold. You will really make that particular effort stick in the dog's mind. Eventually, the dog will start to play you like a slot machine with the idea that maybe this time he'll get a treat, and maybe it will even be a jackpot.

More Than Cookies and Balls

The world is full of positive and negative experiences. Pain and pleasure are the controlling forces of life. It's unrealistic for most people to think they are going to train their dogs without ever correcting them. However, it can be done with enough repetitions, rewards for their good behavior, and a proper leader-follower relationship. Eventually you should only need to give a command and your dog should comply.

The form a correction takes depends on many factors: the dog's temperament, the dog's age, the behavior you are trying to change or correct, the ability of the owner to perform the correction, the dog's understanding of what he has been asked to do or stop doing, and the owner's relationship with the dog. Many corrections are never appropriate. NEVER (this is just a small sample):

- pin your dog on his back to the ground (also known as the alpha roll)
- hit your dog
- whack him with a rolled-up newspaper
- knee him in the chest
- step on his toes
- submerge his head under water to stop him from digging
- shove his nose in pee or poop
- shriek or scream at him
- NEVER PUNISH AFTER THE FACT OR OUT OF ANGER!

These are not corrections; they are dog abuse. Some examples of good, effective corrections are withholding treats for noncompliance; using a low tone of voice in conjunction with an instructive reprimand; using a noise deterrent; and, lastly, if appropriate, using leash corrections. You need to recognize the inappropriate behavior when it occurs and correct it appropriately. Always treat your dog humanely.

Appropriate Corrections

Why you choose to correct your dog's behavior is as important as how and when. Be careful to never overcorrect or undercorrect a behavior. This is where timing and skill are imperative. Obviously, you want the correction to match the behavior. For example, you would never use a muzzle grab to make your dog sit. Likewise, you would not push his rear down to make him stop barking. The punishment must fit the crime as well as the criminal. For instance, I choose to live with Ruby's behavior of jumping up; it's merely a nuisance. But I have used a leash correction to teach her not to run into the street. That behavior would be considered life threatening. Always keep the dog's temperament in mind. An aggressive dog might retaliate, and a timid dog might become terrified beyond repair.

Levels of Corrections (Always Remember: Do No Harm!)

1. Always try to start out with the least you can get away with, meaning hands off.
   - *Withhold rewards:* One of the most powerful corrections is not to reward the dog for noncompliance. If he does not listen to a command, don't give him a treat.
   - *Noise sensitivity:* If the dog ignores your command, repeat it, but add some strength to it by stomping your foot on the ground and saying it in a low, more guttural tone. The dog that is not noise sensitive will ignore you or look at you like you are crazy. On the other hand, be careful not to blow a noise-sensitive dog across the room with the command. You want to get him to stop, not send him to therapy.
   - *Ignore bad behavior:* Wait for the dog to stop, and then reward him for good behavior. Again, this takes a lot of patience, but it can be successfully accomplished.
   - *Time-out:* Put the dog in the crate. Don't think the dog is contemplating the error of his ways. You are preventing him from continuing the behavior at that moment and denying him your attention.

2. When the hands-off approach does not work, try a physical correction. Please note that physical corrections vary depending on the dog's age and temperament and the owner's capabilities.

- *Physical manipulation:* Physically put the dog into position but don't squash him; place him gently. Tuck his rear for sit or hold him down for down-stay. Specific methods of handling need to be used to achieve this, as explained in Chapter 10.
- *Leash corrections:* A short, quick jerk on the leash in the direction you would like the dog to go should always be preceded by a verbal warning or command. In the future, you will only need to warn with the verbal command, and the dog should comply. Do not pull and hold on the leash, strangling the dog; snap it quickly. You want your dog to take you seriously without hurting him.
- *Muzzle grabs and cradle position:* Muzzle grabs and cradling are for puppies five months and under only! They may serve to antagonize a bold, pushy puppy if not done correctly. Again, with a sensitive puppy, you always have to be careful not to overcorrect.
- *Sprays:* For jumping or barking, say "quiet" or "off" and use a breath spray directly sprayed in the dog's mouth. These do not physically harm the dog in any way but do taste most unpleasant to a dog.

3. For certain behaviors, head harnesses, citronella bark collars, citronella remote collars, and citronella boundary collars are extremely humane and highly effective. In worst-case scenarios, pinch collars and electronic bark collars also have their place. When used appropriately—meaning on the right dog, for the correct behavior, and by an experienced person—they are not cruel. Consult a professional trainer or behaviorist before employing these methods by calling 800-PET DOGS (800-738-3647) for a trainer in your area.

Remember, before you employ a correction the dog must first understand the behavior you want from him, which for some reason he's not performing. Maybe he's distracted or preoccupied. Still, the dog has a choice. He can choose to stop what he's doing and obey you, or he can choose to ignore you. Sometimes even a treat is not as interesting as the dog he's playing with. This is where you would bring in the correction. You want the dog to know he will do the behavior for no other reason than because you asked him.

Negative Pairing

It is extremely important to use an instructive reprimand while administering a correction. This way the dog associates the punishment with the command. For example, when the dog does not stop barking, don't just spray breath spray in his mouth, but precede it by telling him "quiet." The dog learns if he is not quiet, he might get the unpleasant taste of breath spray in his mouth.

Don't make the same mistake many of my clients make. They often say to me, "That works great. I only have to show the breath spray to my dog and he becomes quiet." That is not the goal. You don't want to depend on holding up a container of breath spray in order to quiet the dog. The goal is for the word *quiet* to represent the idea of the correction. It's as if you are showing the dog the word, not the breath spray. The word or command alone should be all-powerful.

Again, how you reprimand your dog depends on his temperament. An aggressive dog could very well lash out at you later. If the dog associates a harsh punishment with a command, you might only need to give the dog the command to trigger a retaliation. Be very conscientious of your dog's personality before administering a reprimand.

The Power of a Correction

There is sometimes a place for correction. When used properly, it can be extremely effective. It definitely speeds the training process. At the same time, you don't want to continually administer little nagging reprimands. You will desensitize your dog if he hears "no, no, no" all the time. It is much more effective to use one firm reprimand. You will really get the point across, and it is generally enough to make a lasting impression on the dog.

It's important to understand that punishment does not alter behavior. It merely suppresses behavior, temporarily giving you a chance to get in there and reward good behavior. It provides a fleeting opportunity for you to put power behind the command. But be careful, and consult a professional trainer or behaviorist if you are unsure.

The Training Paradox

None of this is cut-and-dried. The idea that you have to dominate your dog is not what it's all about, and yet you do need to have his respect. For some people, this seems to be a paradox. You need to have a relationship with your dog in which he takes you seriously. This does not mean the dog loses his soul and personality in the mix. I don't want that. I want a dog that shines. While it is important to have the last defining word, you don't want to stifle or stymie the dog with dominating tactics. There is a delicate balance between getting him to want and like doing the behavior and knowing that he will do it because you have a strong leader-follower relationship.

You must start with an understanding of your dog. Based on that understanding, you can train, discipline, and love him with a balanced approach. Why isn't this dog listening to me? Is this a relationship problem or a training issue? We expect our dogs to play by the rules, but we don't always make those rules clear. It takes quite a bit of time and effort to bring an animal into your life and establish communication with him. Sometimes you have to accept and work around your limitations with your dog.

Once you figure out your dog's personality, you will have a better grasp on how to train him. Some dogs are so aggressive you should not use any force whatsoever. It could endanger you as well as escalate the problem. Others are so timid that a reprimand might send them squealing into a corner, creating a bigger disaster. Some people underreprimand their dogs, and the dogs end up dominating the masters.

A dog that is spoiled and out of control needs to have some boundaries. What do you do? You give him boundaries. Keep the dog on a leash, use a crate as needed, or keep the dog by your side during his training period. You need to keep a close eye on your puppy or dog until he is trained. Don't let him get out of your control. Give him actual physical boundaries, and he will learn his limitations. Once the dog is trained, you can give him a little more freedom.

A trained dog realizes his boundaries. When you do not limit a dog's choices, you give the dog too much control. Many people do that with their children. They give them too many choices. It's too many choices for a kid and too many choices for a dog. When you take away the choice and let the dog know he must obey you, you give him structure. If you truly love your dog, give him structure and boundaries. He will feel safe and secure, and as a result, he will flourish.

In Short

The field of dog training has run the gamut from military-influenced, correction-based training to never-say-no, "reward the positive and ignore the negative" motivational training. Drawing from the fields of behavioral science and learning theory, we need to find a happy medium. Teach by using positive reinforcement and rewards. Occasionally a well-timed reprimand will be in order when a dog that knows a command chooses to ignore it.

Dog training boils down to REQUEST + ACTION = CONSEQUENCE. The command is given and the dog's response or action has bearing on the consequence, either positive or negative.

Dogs live in the moment. There must be no punishment after the fact. Dogs should only be corrected when caught in the act of misbehaving. Giving instructive reprimands like "quiet," "off," and "outside" interrupts the misbehavior and gives the dog an alternative.

Dogs do not generalize. They are situational learners. Each command should be practiced under numerous different circumstances.

Most dogs will happily work for food. By saying "good" before giving the food reward, you will create a conditioned reinforcer. "Good" alone becomes meaningful.

A simple, effective way to train a dog is through lure-reward training, where the dog is lured into position with a treat.

By randomizing the rewards and issuing jackpots for a job exceptionally well done, you keep a dog's interest and shape his behavior to your liking.

When faced with a situation in which a correction is in order, remember that corrections must be timely and fit both the crime and the criminal.

Before you resort to a correction, the dog must understand the command. It is imperative to know your dog's limitations. Once you understand the dog, you will have a better grasp on how to train him, for there is no one method that is right for every dog. Give him some boundaries and watch him flourish.

8

Starting Off on the Right Paw

When you bring a dog into your home, you embark on a lifelong journey with him. Everything you do from the minute you bring your dog into your home affects your relationship with him. Get off to a good start by being prepared.

Equipping Your Home

You wouldn't bring a child into your home without the necessary preparations: a place to sleep, food, clothes, toiletries. Why should bringing home a new dog or puppy be any different? Buy your dog's supplies prior to his homecoming. Rather than stumbling around until you find the right equipment, take control and make the proper preparations.

Accessories

When everything your dog requires is already available, you need not scramble around town trying to get organized while worrying about the dog peeing on the rug or chewing on the coffee table. A little forethought can make your life much easier.

Food Stuff

Food items include:

- *Food:* find the most natural, high-quality kibble and canned food available.
- *Bowls:* separate food and water in stainless-steel or ceramic bowls; some dogs are allergic to plastic, which can become scratched and hold bacteria.
- *Treats:* freeze-dried liver or turkey hot dogs are easy to use in training and help promote desired behavior. Dog jerky, biscuits, or cookies should be as natural as possible. Raw carrots help keep teeth clean.

Collars

You won't need all of these, but here's what is available:

- Buckle collars are for all ages; they are used to attach ID tags and as the primary training collar.
- Nylon slip collars are for training; they follow the same principle as a choke chain but are made of nylon.
- No-pull harnesses are anti-pulling devices; they are good for puppies, smaller dogs, and medium dogs.
- Head halters are for more boisterous, out-of-control dogs; shy dogs; aggressive dogs; and dogs that chase and pull.
- Pinch collars are to be used on *rare occasions* with the help of a professional.

Leashes

Available leashes include the following:

- Six-foot leashes are for general training.
- Four-foot leashes are for controlling the dog's movement in the house.

- Fifteen- to twenty-foot leashes are for teaching obedience at a distance and working toward off-leash control.
- Puppy leashes are extremely lightweight, thin, and nylon; they do not bother the puppy too much. Cut off the handles so they can't catch on anything.
- Tab-short leashes are for corrections and guiding the dog.

Chew Toys

Please buy toys that are appropriate for your dog's size. Here's what is available:

- sterile bones you can stuff with jerky treats
- round flip chips bigger than the dog's mouth
- pressed rawhide—avoid particle rawhide, which falls apart
- Kong toys, which are hollow, rubber, and easy to stuff
- Buster cubes, which are hollow and dispense treats and food
- raw cow bones—knuckle and femur bones are best; cooking causes them to splinter

Hygiene

Hygiene items include:

- a brush that suits your dog's coat; ask the vet, groomer, or breeder
- nail clippers
- a styptic stick or powder to stop bleeding in case the quick is nicked when clipping nails
- shampoo and possibly flea products—ask your vet
- a flea comb

Essential Extras

You should also stock up on these essential extras:

- anti-chew cream and spray—the cream is used for hard objects like table legs, walls, doors, and doorjambs; the spray is used for soft, clothlike objects like leashes, carpet fringes, or shoelaces; once the spray dries, however, it is no longer effective
- human breath spray to help teach the off and quiet commands
- odor neutralizer to clean potty accidents
- an ID tag with your name and number, which should be immediately placed on the dog or puppy; there is no age requirement for an ID tag
- a dog license tag—check with the local authorities for your community's licensing regulations
- an exercise pen—a virtual puppy corral; collapsible, moldable, and can become an indoor sanctuary for the dog; also can be used to block off a portion of your home

Dog in Sherpa Bag ready to go

- a crate for housebreaking, time-outs, and preventing destructive behavior
- a bed—not necessarily good for puppies, which may eliminate and chew on them
- a puppy gate to block off rooms
- a Sherpa bag—great to carry small dogs; fits under an airplane seat

A Note on Leashes

You can never have too many leashes, but you can have inappropriate ones. Don't anchor a young puppy with a huge leash. Leash width should be age appropriate. Puppies should have a ¼-inch-wide or smaller leash, depending on their breed. Adolescents and adults should have a ½-inch to a ¾-inch leash. The most comfortable leashes are leather. Nylon is acceptable, especially for puppies, which may go through several leashes.

Snaps

There are two types of snaps, the metal piece on a leash that attaches to the collar. A bolt snap has a metal eye that is opened vertically by pulling down on a small metal thumb piece. A spring snap is pressed inward to open it. I prefer bolt snaps. Spring snaps can inadvertently open from any slight pressure, freeing the dog. The dog's toes or other body parts can also get caught in the snap. Spring snaps "snap" shut.

Retracting Leads

Stay away from retracting leads unless you are all by yourself in a big field. The long, thin line easily cuts through skin, causes burns, and tangles and trips both people and dogs. One of my clients lost the tip of a finger when her dog abruptly lunged and pulled on the line. If you drop the large plastic handle, it bounces after the dog, often scaring him. The dog may try to run from it into dangerous situations like traffic.

Choke Chain Alternatives

I generally find choke chains ineffective and too difficult for most dog owners to use. Most dog owners never master the complex combination of skill, timing, and force needed to make a proper correction with a choke chain. As a result, dogs may be reprimanded incorrectly and too harshly. Even if a dog owner properly uses a choke chain, the dog may not respond. Some dogs are more able to tolerate the pain incurred from choke chains.

No-pull harness prevents pulling when walking.

For these reasons, I find no-pull harnesses and head halters much more effective and humane. They are similar to the equipment used on other animals such as horses. Make sure you use a no-pull harness because regular harnesses actually foster pulling. Despite their muzzlelike appearance, head halters allow the dog to open his mouth, eat, and pant normally. Head halters are designed around a dog's natural movement. They tug at the back of the dog's neck much like a mother dog holds her puppies, and they loop around the dog's muzzle, containing it much like a muzzle grab. Head halters have a calming effect on many dogs and are used for pulling, boisterous, out-of-control, fearful, or aggressive dogs. Most dogs may protest at first, but they quickly learn to accept their head halters. Head halters and no-pull harnesses are usually available through dog supply catalogues.

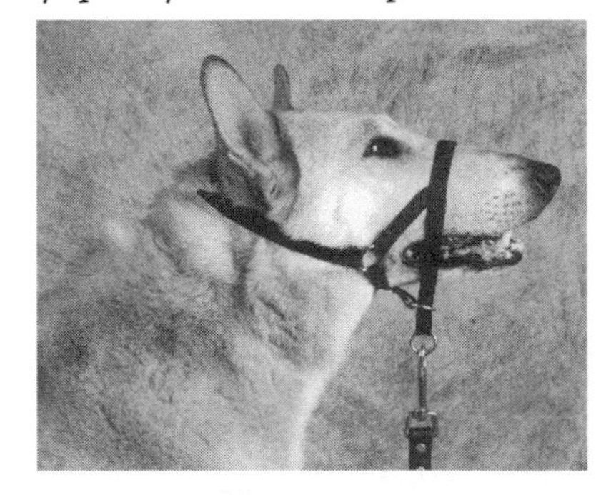

Head halters allow the dog to open his mouth normally.

Dogs are the only domesticated animals on which we use metal collars to control their movement. I used to stop people on the street who used pinch collars on their dogs and lecture them on what I called "the medieval torture device." Although they are rarely needed, I have since come to appreciate the effectiveness of pinch collars used on the right dog for the right behavior. Despite their menacing look, pinch collars are kinder than choke chains because they get the point across quickly and without strangling the dog. Pinch collars give you power steering over your dog. To see how they feel, try putting one around your arm. You will find it is not as bad as it looks. Pinch collars are never to be used on fearful, shy, or aggressive dogs and should only be used under the supervision of a professional trainer.

Ideally, you should train your dog using a buckle collar, but many dogs will require other equipment for specific training needs. Use other types of collars until you overcome the problem, and then revert back to using a buckle collar.

Fitting Buckle Collars

A properly fitting buckle collar allows a two-finger width between the collar and the dog and cannot be pulled over the dog's head and ears. Measure your dog's neck and add approximately two inches. Like leashes, collars have two types of buckles. The touch-lock clasp is made of plastic and can be cracked, broken, or bitten through. The best collar to buy has a metal buckle. Buckle collars have been around for years and stand the test of time.

Leather vs. Nylon

Leather can be expensive. You may want to wait until the dog reaches his adult size before purchasing leather leashes and collars. Leather feels good, is attractive, lies nicely in the hand, and has a good amount of give-and-take in terms of control. Avoid chain or metal leashes. Metal does not conduct physical messages like leather does. Sure, metal is durable, but you cannot feel and judge your dog's movement as well with it. Also, metal is heavy and can cause pain. Don't buy metal because your dog chews his leash. Instead, address the problem, and teach him not to chew his leash.

Crate Training

The crate is the most useful piece of equipment you can buy for your dog. It is also one of the most misunderstood. Dogs are den animals. They prefer small, confined spaces. Humans tend to view confinement as claustrophobic and imprisoning. Herein lies the problem most people have with crate training. Please understand that a crate in your home takes the place of a den in the wild. Crates are not cruel and unusual punishment. In fact, they are quite the opposite. A crate is a haven and a refuge for your dog, making him feel safe and secure.

I understand the hesitation to crate train. When I first brought Lotte home, I put her in a crate. She cried bloody murder, and I immediately took her out. I disassembled the crate and threw it in the closet.

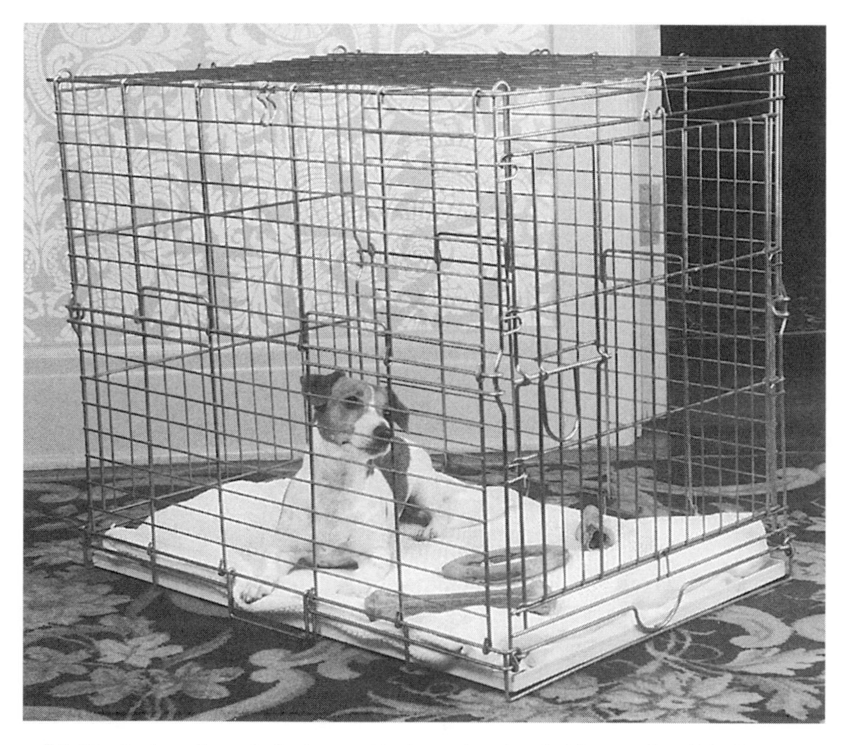

A happy dog in a wire crate lined with a blanket and equipped with pressed rawhide chew toys and a stuffed sterile bone

Ironically, a year later when we started competing, I had to put Lotte in a crate to attend shows. I put her bed in the crate, and Lotte walked right in and went to sleep. I was lucky. Crating later in life can sometimes be difficult for your dog. This is why it is important to accustom your dog to his crate right from the start. At some point in his life he will have to go into a crate, whether for travel, boarding, or grooming, or visiting the vet.

Clients frequently tell me they have a crate, but the dog does not like it so they don't use it. Meanwhile, the dog runs loose around the house wreaking havoc, or the owners construct their own pseudocrate. Some people go to great lengths to make a denlike space for their dogs. They pen the dog in a small space or construct boxed areas with high walls. Even though a crate is almost the exact thing they have built, people refuse to use a crate because of the stigma it carries. It's ridiculous. A crate is so simple and self-contained. It has a top and a bottom so the possibility of the dog getting out is not a problem, and it prevents any urine or feces from touching your floor if an accident does happen.

Why Crate?

The crate serves several purposes. Housebreaking takes top billing. Most dogs are by nature very clean animals. They do not soil where they hang out. Housebreaking is about prevention. The crate is a safe place to keep your puppy when you cannot watch him. If you do not allow your dog a first potty accident, you are off to a great start. Confining the dog to a crate is the best way to prevent an accident because it enables you to control his potty schedule. This helps you judge when your dog needs to go potty so that you can take him to an appropriate area. Crating also helps teach your dog bladder and bowel control.

The crate is also used to prevent destructive behaviors like chewing or digging. If your dog is in his crate, he cannot chew the rug or dig up the backyard. Again, *prevention* is the operative word. If a puppy never partakes in destructive behaviors, he will be very unlikely to participate in them when he is older. You prevent the behavior from happening or continuing when you crate your dog. When he is crated, give your dog an appropriate chew toy, teaching him a proper chew-toy habit.

Dogs feel safe in their crates. Once they become accustomed to their crates, most dogs love them. The crate provides a dog with a sense of safety and security. Dogs naturally seek out small, enclosed areas. You will often find your dog sleeping under a bed, a table, a chair, or in a corner.

Dogs are not only den animals but pack animals as well. In the wild, a dog is never left alone. When a dog separates from his pack, his natural instinct kicks in, and he cries for the rest of the pack. This is called a separation reflex. It is the reason your dog or puppy cries for the first few nights after you bring him home; he misses his previous pack. If you coddle him, his crying will persist when you are gone. Putting him in a crate where you give him treats and try to make the experience pleasant eventually eases his anxiety and builds his confidence. He quickly learns that he is safe in the crate and you are his new pack.

Types of Crates

There are two types of crates. Each has its advantages. Choose a crate based on your lifestyle and use of the crate. One type is made of plastic and can be used to transport a dog on a plane. This is a good choice for a frequent traveler. It tends to be less expensive, and many dogs prefer the security from the enclosed feeling of the plastic crate.

The other option is a wire fold-down crate. This is my preference. It is cooler in the summer and easily folded for traveling. Potty accidents are easier to see through wire crates, too. To make a wire crate more den-like, drape a towel or blanket over the top of it.

Crate Size

The crate should be large enough for your dog to stand and turn around in but small enough to inhibit him from soiling. Buy a crate for the adult size of your dog. If you get a puppy that will grow into a large dog, buy a large crate. Do not buy a small crate, because your puppy will outgrow it the same way a child outgrows a pair of shoes.

Because you buy a crate for your puppy's adult size, you will need to shrink the available room in the crate to prevent soiling. Some pet stores sell crate dividers for the wire fold-down crates. A hard, plastic cooler shoved to the back of the crate works well, too. Get creative, but make sure whatever you use is sturdy enough to withstand a feisty puppy. As the dog grows, increase the amount of room in the crate. He should always have enough room to stand, turn around, and lie down comfortably.

Equipping the Crate

Initially, there is no need to buy an expensive crate pad or blanket for the bottom of the crate. A towel is sufficient and easily washed. An extremely young puppy or a dirty puppy, which is a puppy that has learned to soil where he sleeps, may soil in his crate. If this is the case, take away the towel. Once the habit is broken, you can experiment with adding towels.

Always put an appropriate chew toy in the crate. The toy keeps the dog or puppy busy when he is not sleeping. When he has a good chew toy to keep him occupied, your dog will not be so interested in getting out and will not bark for your attention. It's also a good way to start your dog on a proper chew-toy habit. It's better that he chew a toy than the fringe on your rug.

How to Crate Train

Some crate-training methods tell dog owners to never reprimand their dog or do anything negative to him while he is in the crate. It is also thought that dog owners should never force the dog into his crate or put him there after he has engaged in inappropriate behavior because it is believed this causes a negative association with the crate. I generally find these notions untrue.

Into the Crate

To get your dog or puppy into his crate, lure him in with a treat. Toss the treat in the crate and use a command like "go to your crate" or "go to your bed." If he does not go in voluntarily, pick him up, place him in the crate, and shut the door. Once he is in, make up with him for pushing him in by giving him a treat and saying "good crate" or "good bed."

I do not tolerate a dog that does not want to go into his crate. While it is great to lure the dog into the crate voluntarily, I find that at the slightest hesitation it is better to pick up the dog and gently put him in the crate. The next time, the dog is more likely to go right into the crate because he learns balking will not work. Of course, you would handle an aggressive dog much more delicately. If he really resists, do not force him.

The Release

Do not make a fuss when you let your dog out of the crate; otherwise you will be rewarding freedom. Instead, ignore him for a while, or act nonchalantly before you give him attention. Act this way every time you let him out of the crate. Do not, under any circumstances, let your dog out of the crate when he cries, paws, or protests. He will interpret the release as a reward for throwing a tantrum. Wait until he is quiet and calm, even if for a few seconds, before you let him out.

Reprimands in the Crate

When a dog barks incessantly while crated, you have two choices: you can ignore the dog, or you can make him stop. My experience is that the fastest, most efficient way to end the behavior is with a reprimand. Some people may disagree with this. They say the crate is the dog's safe place, and nothing negative should ever be done to him while he is crated.

I do not tolerate crate protests. The faster you eliminate the protest, the sooner he begins to love and accept his crate. It's fine to ignore slight whimpering when barking is the major issue. Once you control the barking, then deal with the whimpering. Before you reprimand your dog for barking in his crate, give him the benefit of the doubt and make sure he does not need to go potty.

As with any reprimand, I start with the least harsh. In a low, guttural voice give the instructive reprimand "quiet" to a crated, barking dog. If the barking persists, tell him "quiet" again, and firmly tap the side or the top of the crate. The dog should stop at this point. If he continues, try saying "quiet" again, and spray him directly in the mouth with breath spray. With a young puppy that barks relentlessly, you might try a muzzle grab.

When you catch your dog in the middle of an undesired behavior like digging or chewing and you cannot actively work on changing the behavior, it's okay to give him a time-out and put him in his crate. Don't be fooled and think your dog will contemplate the error of his ways during his cooldown in the crate. Dogs don't think like that. He won't dislike his crate or think he is being punished if you send him there. He probably prefers to be in the crate when you are in a frenzy because he knows you can't get to him. Crating your dog prevents him from continuing the destructive behavior. Put a chew toy in with him to make him happy and to constructively drain some of his energy.

When to Crate

Crate your dog:

- when he is tired
- at feeding time
- at night
- when you cannot watch an untrained dog or puppy

Puppies sleep a lot. Let them! Do not crate your dog or puppy when he is most active. Let him play, or better still, exercise with him. Once he has eliminated and he is tired, then crate him. Never crate a full dog. This may seem to contradict crating him at feeding time. Let your dog or puppy eat in the crate, and then take him out to go potty. Don't keep him in the crate more than fifteen minutes after he finishes eating. Some puppies need to eliminate right away. Feeding your dog in his crate is one way to create a positive association with his crate.

During the day, depending on his age, a puppy can be left in the crate for two to three hours at a time and eventually up to four hours. This frees your schedule. Training a puppy does not mean your life stops. Crate training is also only temporary. How long you crate the dog depends on the individual dog. Sometimes as little as six months is sufficient, but that does not see your dog through adolescence, a time of behavioral inconsistency. More likely, you will need to crate your dog until he is one to two years old. I crated Ruby for a little over two years because I knew that her potential for destruction was high. Now she roams free through the house, and I have incurred no damage because of her.

Overcoming the Separation Reflex

A puppy or dog will most likely cry the first few times he is left alone and crated. Often, it is his separation reflex at work. He yearns for his previous pack or to be close to you. It is your job to ease his transition and help him accept his crate. The worst thing you could do is ostracize him in a laundry room or bathroom in an attempt to potty train him or keep him from destroying your home. This is where a crate proves most useful.

Crate your dog in a heavily trafficked room so that he can see everyone and feel a part of the family. At night put the dog's crate in your bedroom next to your bed. Ideally, in the beginning he should be able to see you and make eye contact with you. You may even stroke your dog or puppy through the crate to comfort him. A young puppy will particularly appreciate the tactile sensation since he is used to the close feel of his mother and littermates. Don't coddle him, and don't let him get too dependent on your stroking. He may quickly learn that a whimper is a good way to get attention. Whatever you do, do not take him out of the crate except for those middle-of-the-night toilet trips you will need to make with a young puppy. Your dog can only overcome his anxiety by learning that he is with a new pack and is fine when left alone.

Training Your Dog to Be Alone

To successfully live with humans, dogs must learn how to feel relaxed when left alone. The crate is the perfect tool to teach this. Start by crating the dog in one room while you are in another. Walk in and out of the room in which your dog is crated, showing him you return after you leave and there is no cause for alarm. Systematically begin to leave him for longer periods of time. You might leave the house for twenty to thirty minutes at a time. When you return, don't let him out of the crate right away or fuss over him. By acting nonchalantly, you help keep him calm. Gradually increase your time apart. You should be able to leave the dog for up to three hours, depending on his age.

Dogs and Children

When we think of dogs and children, we often imagine a scene of the two frolicking together on a lush lawn, but reality is often not the beautiful picture we conjure up. Many dogs do not know what to make of children, often viewing them as littermates or play toys. A child's smaller size and sometimes awkward movement can be strange and alarming to dogs. Also, a baby's cry or a child's scream is similar to the sound of a puppy or wounded animal—noises to which a dog is apt to respond. More than two million children are bitten by dogs each year, and most of these cases occur with the child's own dog or that of a friend, neighbor, or relative. This is one reason children under twelve should not be left unsupervised with a dog.

Parents need to see themselves as mediators between their children and their dogs by making the dog feel comfortable around children and by teaching children how to interact with dogs. There are steps we can take to keep the peace between our dogs and our children:

- Ensure your leadership position over the dog by making him work for food, praise, etc.
- Teach your child how to touch and handle your puppy or new dog. Children under seven should never be allowed to pick up the dog since they are usually awkward and attempt to do so by the forelegs, leaving the rest of the body hanging.
- Obedience train your dog and practice with him around children so that he learns to listen to you when children are present.

- Do not roughhouse with your dog, play tug-of-war, chase, or wrestle—these games encourage the dog to play rough with children.
- Teach children appropriate games to play with dogs, like fetch and hide-and-seek.
- Do not use overly dominant training techniques with the dog—he may learn to obey you but may treat anything smaller than himself, like a baby or child, aggressively.
- Keep your dog on a leash and possibly even use a head halter on him until he is comfortable in the presence of children.
- Above all, teach your children to love and respect the dog; they should be kind and gentle to the dog.
- Always supervise your children and dog when they interact.

The Child's Role

Many parents get a dog as a way to teach their children responsibility. Even in families with children over twelve, it may be unrealistic to expect children to take full responsibility for the dog. The child should participate in the training and care of the dog but should not be solely responsible. A child who is involved with the dog's obedience training usually has more confidence in the ability to give a command and get compliance. However, children should never perform a muzzle grab because they have a hard time asserting enough dominance over the dog and can often antagonize the dog. Reprimands are okay only if backed by an adult repeating the reprimand. When an adult supports the child's reprimand, he helps transfer authority to the child so that the dog takes the child's reprimand more seriously. Kids must also learn not to run or roll around the dog and to maintain a higher physical stance. When a child's posture is heightened around the dog, chances are that the dog will see the child as more of a leader.

Possessive dogs can be problematic when a child or toddler nears the dog's bowl or toys. Combat this situation with food-bowl work and systematic desensitization from day one. Practice by putting your hands in the dog's food bowl, giving him treats as your hand approaches the bowl. The dog begins to associate a hand by his bowl with something positive like a treat. This must be done very slowly and sensitively. If your dog shows discomfort at any time, back off to a point where he is more relaxed. Ease into it. You can also trade the dog a treat for his bone or toy. Teach him the give command, and when he hands over his coveted item, immediately give him a treat.

Introducing Dog to Baby

Many couples are waiting to have children. They often get dogs in the interim as child substitutes. By the time they do decide to have a baby, the dog is well established in the home. Many owners wait until the baby is born before exposing the dog to a child. This can be problematic, and most people will want to do more. The best possible scenario is one in which the dog has already been exposed to children in a positive manner, making the transition much easier. He should be given obedience commands and treats in the presence of children in order to make him more manageable and more at ease around them.

Another way to prepare the dog for your new family member is to acclimate him to baby items such as baby powder and receiving blankets. Wrap a receiving blanket around a doll, and allow the dog to investigate as long as he keeps all four paws on the ground. If you need to make a correction, put the doll down to avoid a negative association between the doll and the reprimand. Whenever you hold the doll, and later the baby, teach the dog to sit or lie down. This will help transfer your dominance over the dog to the baby. Also, begin to introduce the carrier and stroller on walks before bringing home the baby so that the dog can get used to walking this way.

Because dogs are naturally attracted to the sound of cries, your dog will probably want to run to the source of the sound to investigate. Acclimate your dog to baby cries before the baby comes by bringing home a tape of a baby crying. Put the tape recorder in the baby's crib, put the dog on a leash, and teach him to come to you whenever the cry is heard.

Teaching the dog to come to you rather than go to the crib when the baby cries averts a potentially dangerous situation and teaches the dog to notify you when the baby is crying.

When you bring the baby home, the dog should be introduced to a piece of the baby's clothing or some other item carrying the baby's scent before the baby enters the house. The mother should enter first while the baby is held outside. Once the dog satisfies his greeting ritual and is calm, the baby can come into the house. A dominant dog should initially meet the baby off the property and the baby should enter the house first in order to establish dominance order. The baby should always be kept at a higher level than the dog, both literally and figuratively.

Many people mistakenly ignore or ostracize the dog when dealing with the baby and only give him attention when the baby is napping. The dog begins to associate good times with the absence of the baby. Instead, you should ignore the dog when the baby is napping and lavish your dog with love, attention, and treats when the baby is present. Try to make the dog as comfortable as possible around the baby. One word of caution: never leave the dog alone with the baby, and never put the baby on the floor in the presence of the dog.

Before Bringing Home Baby

When singer Melissa Etheridge and her companion, Julie Cypher, decided to have a baby, they called me to help prepare their two dogs, Bingo and Angel. Both dogs were a bit uncomfortable around children, Bingo more so than Angel. We put a head halter on Bingo to give us more control. We began by slowly introducing the dogs to children of various ages while using treats. We started at distances comfortable for the dogs and fed them treats whenever a child was visible. As soon as the child moved away, we took the treats away. Soon, both dogs began to associate children with something positive—treats.

Next, when the dogs were more comfortable, we practiced obedience commands in the presence of children. Whenever they would see a child, we told them to sit. Gradually, we could get them to interact with children. We had children drop treats on the ground for the dogs to pick up. Once the dogs were extremely comfortable, we allowed children to feed them treats from an open hand. Never at any time should the child reach toward the dog. Rather, the child should wait for the dog to approach. The dog should also be on a leash and wearing a head halter at first. Eventually, Melissa and Julie's dogs were more comfortable around children, making the introduction of a baby into their home much easier. Happily, on a return visit, I found the dogs comfortably interacting with the baby.

It can be helpful to begin working your dog around children on neutral territory. The dog will probably be less territorial off your property. Also, never force the dog toward the object of his fear. Every step must be slow and small. Pay attention to your dog, taking your cues from him. If he becomes fearful or agitated, stop. It's a good idea to seek the help of a professional trainer or animal behaviorist when doing this type of work with your dog.

Diet and Exercise

While obesity can impose a threat to your dog's health, there are other reasons to maintain your dog's diet and exercise regime. Diet and exercise can markedly affect your dog's personality and behavior. You want to choose a diet regime that is most natural for your dog.

Protein Conscious

Many puppies, particularly larger breeds, receive too many nutrients, including protein, in their diets. Many breeders and vets are now placing puppies on adult dog diets. Dog food labeled for adults has less protein than that labeled for puppies. Oversupplementation is thought to foster conditions like hip dysplasia. A high-protein diet may increase aggression and activity in all dogs. If you've got a hyperactive or aggressive dog, consider trying a low-protein diet. Dog food labeled for senior dogs is lowest in protein.

Haunches of Steel

The majority of dogs do not get enough exercise. Ideally, most breeds should have a strong aerobic workout for at least forty-five minutes to one hour twice a day. In addition, twenty- to thirty-minute walks twice a day are also great exercise. Daily walks coupled with two weekly big bursts of exercise until the dog is out of breath is an acceptable exercise program. Exercise is age and breed specific. An older dog may only be able to handle daily walks. More heavy, low-slung dogs like Basset Hounds may not have the stamina for an hour of high-energy exercise. Smaller dogs may get all the exercise they need from running around the house or playing fetch, but should still be walked for the mental stimulation.

Start an exercise program gradually until you build your dog's or puppy's aerobic capacity. Never exercise him shortly before or directly after he eats. When it's hot out, be mindful of your dog's breathing—especially with breeds like Bulldogs or Pugs, who may have touble breathing to begin with. If he pants heavily, slows down, or begins to act strangely, stop and let him cool down. Heat exhaustion can be dangerous if not caught in time. Also, be mindful of the surface on which you exercise him. Pounding exercise like jogging or running should be done on grass or dirt, not hard pavement. This is especially true for a young dog that is still developing orthopedically. When the weather is hot, avoid blacktop and asphalt, which can get very uncomfortable.

Gentling and Handling

At puppy class graduation we dress the puppies in costumes. This is not just to be cute. If your dog accepts being dressed up, he will accept much of the touching and handling required throughout his life. Many people are not able to brush their dogs. At best, the dog squirms and tries to get away. At worst, the dog bites the brush or even the owner. You must be able to touch your dog anywhere on his body so that you can groom him, take debris out of his coat, clip his nails, check his mouth, clean his ears, or dress him with bandages. The dog must be comfortable when being handled.

With all handling exercises, start at a point the dog accepts. Use positive reinforcement by giving him treats when he lets you handle him. Systematically work your way to trickier spots on his body. Many dogs need to be anesthetized for simple procedures like ear cleaning. The goal is to be able to poke and prod the dog anywhere.

If you have a fearful or aggressive dog that resists any of the gentling and handling exercises, change the way he feels about it through systematic desensitization. Start lightly and then build. Do not use force. Start by merely showing him the brush or clippers or touching him on a spot he accepts. Reward him with treats when he does not react. Gradually increase contact, continually rewarding him with treats for every bit of success. Eventually you will be able to brush him or stick your fingers in his mouth without worrying or frightening him.

Treats

The best way to ease your dog through an unpleasant experience is using treats in conjunction with verbal praise. If at any time during a handling exercise your dog shows signs of distress or aggression, back off. When you make any progress, reward him with a treat. He learns when he lets you touch his body that he gets something he likes. Eventually, the experience becomes pleasant for him.

Likewise when you take him to the vet or give him a bath, have plenty of treats on hand. Give him a treat while the vet sticks a needle in him. Just sitting in a vet's office may be enough of a reason to give him a treat. And baths . . . most dogs dread baths. Keep a bowl of treats nearby where you can easily grab one and reward him for tolerating the soap and water. The idea is to give your dog something pleasant to associate with an otherwise unpleasant experience.

Holding a Puppy

The cradle position is one of the best ways to make your puppy comfortable with being touched and held. It gives you a tremendous amount of leverage when it comes to gentling and handling. Initially, many puppies panic and struggle. If your puppy squirms or resists, hold him tighter. Do not put him down. You must be the winner in this battle of wills. Letting him down tells him throwing a tantrum is a good way to get what he wants. As soon as he calms, comfort him with treats and soft caresses. If he tries to bite you while in the cradle position, perform a muzzle grab if age appropriate (see Chapter 3), or use a head halter.

Brushing

Brush and comb your dog or puppy from the start. It promotes good hygiene, keeps long coats from getting matted and tangled, removes excess hair and dirt, and stimulates oil glands that keep your dog's coat healthy. It is a lifelong routine that should be done on a daily basis. Brushing your dog gives you a good opportunity to check for skin irritations, cuts, ticks, fleas, lumps, or bumps. To brush:

1. Hold a small puppy in a cradle position, or put a dog on the floor.
2. Start brushing his chest.
3. Turn him over in your lap or on the floor, holding him at the collar.
4. Gently stroke the brush down his back.
5. Reward him with treats and praise throughout.
6. Work toward his head.

If he resists, back off and alternate between brush strokes and treats until you are able to brush his head and all other body parts. You can be more insistent with a submissive puppy.

Foot Massage

Foot massage is another gentling and handling habit that should start as soon as possible and continue for the lifetime of the dog. This can be tricky because dogs like to have all four paws on the ground. Treats, along with praise, are the key to mastering foot massages. You need to be able to examine your dog's paws in case he steps on something, develops a limp, or simply needs to have his monthly manicure. To give your dog a foot massage:

1. Gently hold your dog's paw.

2. Touch the top and bottom of the paw.
3. Apply slight pressure as you advance.
4. Give him treats and praise with every step.
5. Repeat the process on each paw.

The Mouth

Touching your dog's mouth is easier to accomplish when you start while he is a puppy. Be careful with an older dog with stronger teeth and a more powerful jaw. It's important to get into your dog's mouth if he is choking, needs to take a pill, or has something stuck in his mouth. Preferably hold a puppy in a cradle position. The dog or puppy should be positioned so he cannot pull back or get away when doing this exercise. Holding his collar at the back of the neck helps keep him still. To examine your dog's mouth:

1. Lift your dog's upper lip.
2. Put your finger in his mouth.
3. Massage his gums and teeth with his mouth clenched.
4. As you progress, open his mouth and massage his inner mouth.
5. Praise him and reward him with treats at every step.

The Ear

Many dogs have floppy ears, which are prone to infection. It is a good idea to systematically desensitize your dog to having his ears touched. He may face a lifetime of ear troubles, which will require much handling. To check the ear:

1. Massage the ear at its base.
2. Gently move toward the tip.

3. Softly move your finger on the underside of the ear, being careful not to go too deep.

4. Continually give him treats.

Prepare Your Dog for Surprises

Gently pulling on your dog or puppy's tail or sides helps prepare him for surprises. A dog's natural reaction is to whip around and very possibly snap at whatever is pulling at him. Your goal is to tame his aggressive, knee-jerk reaction and replace it with a calmer, more gentle reaction. To do this:

1. Start by very softly tugging on your dog's tail or sides so that he only looks back.

2. When he turns to investigate, give him a treat.

3. Tug a little harder to elicit the same response.

4. Reward him with a treat accordingly.

5. The goal is to get him to nonchalantly respond to you without being aggressive.

Body Massage

After a long period of exercise or anytime he's relaxed, your dog would probably appreciate a massage. Wouldn't you? Massage his entire body with large, circular motions. If he's reluctant at first, don't worry. He will get used to it and will learn to enjoy it. You will hear him sigh deeply as he relaxes into your loving touch.

In Short

You can make your first days with your new pet easier if you equip your home and shop for necessary supplies before the new arrival makes his appearance.

Crate train your puppy from day one. As den animals, most dogs will feel comfortable in a cozy, confined space. However, some puppies may need a little time to get acclimated to a crate. Do not confuse initial fussing by the puppy as a sign that he does not like his crate.

To get the dog into the crate, toss in a treat and issue a kennel command such as "go to your crate." If the dog balks, place him inside. This matter is not up for discussion. You can let him out once he settles down for a few minutes. Never open the door when he is in the midst of a tantrum.

A few crating rules:

- Create a positive association with the crate by feeding the puppy while he's in the crate and placing treats and chew toys in it.
- When the dog barks or fusses, either ignore him or give the instructive reprimand "quiet" prior to firmly tapping the crate or using a spritz of breath spray.
- During the day, crate a puppy for up to two or three hours at a time, depending on his age.

When dogs and children are together, an adult must be present to act as mediator. Both adults and children should never roughhouse or play aggressive games like tug-of-war, wrestling, or chase with the dog. Food-bowl work helps combat possessiveness in a dog when children are around. Acquaint your dog with children before you bring your new baby into your home.

Proper diet and adequate amounts of daily exercise are crucial to your dog's healthy development. The proper amounts of each are unique to your individual dog.

Gentling and handling exercises acclimate your dog to the various kinds of touching and handling he will experience in his lifetime.

9

Housebreaking

Housebreaking is about two things: preventing the dog from making a mistake and praising him like gold is pouring from his body when he goes in an appropriate elimination area. Housebreaking is actually very simple. No matter how crazy and out of control a situation may seem, it is always cured by preventing undesired behavior and rewarding appropriate behavior. Interpreted, housebreaking is about what *you* do, not what your dog does. If you do not allow your dog or puppy to have an accident in the house, he starts to respect that area as his living space.

By frequently taking your dog outside to an appropriate elimination area, you teach him where he should go potty, and you help him view the house as a living space, not for soiling. Schedule his meals and water so you have a better idea when he will have to eliminate. Watch everything he does, and when you can't watch him, put him in a crate. You must be completely immersed and involved with your dog or puppy during the housebreaking period, which includes teaching elimination, or "going potty," on command. Finally, when you take him out, you need to physically be there to let him know how brilliant he is for going in the right location.

Creatures of Habit

You are on your way home with your new puppy cradled in your arms. You walk through the front door and place the puppy on the floor, welcoming him to his new home. The puppy sniffs around a bit, looks at you, and pees right there on the spot. Guess what? Your new puppy just formed a habit.

Dogs are creatures of habit, and it usually only takes one incident to create a habit. When it comes to housebreaking your dog or puppy, you need to be aware of his every move. It is much easier to prevent a behavior from happening than it is to break a habit. Be on top of it, especially in the beginning. When does the housebreaking process begin? As soon as you bring the dog home, right? Wrong. Housebreaking begins before you even get your dog or puppy.

Be Prepared

By preparing ahead, you can prevent a housebreaking nightmare. Before you bring home your new puppy, decide where you want him to go potty. The minute you get home, take the puppy directly to the potty spot. My most successful clients are those who consult me prior to getting their dog or puppy.

You will also want to have your house ready for the dog or puppy. Equip yourself with all the necessary tools of the trade: leashes, a crate, collars. By having these items already on hand, you will avoid the stress of worrying about your puppy going potty while you try to get it all together. Make every possible effort to avoid that first potty mistake. You must be proactive when it comes to housebreaking your dog or puppy.

The first thing a new puppy will want and need to do when you bring him home is go potty. He may even pee in the car on the way home, so bring a towel just in case. The movement and excitement of being in the car, possibly for the first time, could stimulate the puppy's bladder. Holding the puppy during the car ride is a good idea because it not only helps comfort him but may also inhibit him from letting go in the car. By the time you get home, you will have a puppy that really needs to go potty. You'd better have a potty spot picked out.

Substances

Dogs develop preferences for certain substances, including your carpet, tile, sisal rug, concrete, grass, dirt, or ivy. A dog that learns to pee on grass may acquire a preference for grass. When I showed Lotte in the Gaines Classic Invitational, which is the highest level of obedience competition, I found myself back in Las Vegas. The motel I stayed at was completely surrounded by concrete. I took Lotte outside several times, utilizing my potty command.

Much to my dismay, Lotte refused to go potty on the concrete. As we left Las Vegas, I stopped at the rather large lawn of a hotel. Being a grass girl, Lotte readily eliminated. This illustrates how faithful a dog becomes to a substance. What does this mean to us? Think twice before you train your dog to go potty on your patio tiles, which are similar to the linoleum in your kitchen.

Location, Location, Location

Unfortunately, teaching a dog to eliminate in a certain area of your yard is not easy. It requires you to escort the dog to a location you deem appropriate. You can hope it is also an area that the dog thinks is a good idea. Your most successful place is an area in which the dog has already eliminated. Then your chances of success are extremely high. Once a puppy finds a location he likes, he continues to use that spot. Think about your first day in school when you picked a seat. The next day, you sat there again, and you continued to sit there until the teacher made you move. It is your job to familiarize your dog or puppy with a proper elimination area.

Be realistic in your potty spot choice. I have a dog run, which is a grass or gravel fenced-in area in the yard specifically for a dog. My dogs only eliminate there because I do not give them any other area. It's quite a simple location for me to train them to go. Stay away from labyrinth-like or foreign areas. I had a client with two Golden Retriever puppies. The client wanted me to train these dogs to pee and poop on top of a drain in the ground that was behind the carport, past the garage, and through a little narrow passageway. It was hard enough for *me* to figure out what the client wanted, let alone a dog. We chose a lovely area in the garden instead.

Situational Learners

Dogs are situational learners. If they learn a behavior in a particular way and place, then that is how and where they know the behavior. This is one of the reasons your dog or puppy should be on a leash when you take him to go potty. If you never take him out on a leash and then one day you do, he probably will not eliminate. He will look at you as if to say, "I don't pee while on a leash." It is also a good idea to take your dog to the chosen potty spot on a leash because you have more control over your dog or puppy. Otherwise, you may be the only one standing by the potty spot.

A Loaded Gun

If your dog does not go potty, don't assume that he is empty. If you do, when you bring him back inside the chances are high that he still has to go and will do so in your house. It is better to assume that he has to go. If he does not go potty, he is like a loaded gun with the potential to go off at any moment. Many people are upset when they send their dog outside and the dog urinates upon returning to the house. These people falsely assumed their dog went potty simply because he was outside. Keep track of when your dog eliminates.

You should always accompany your dog or puppy to the chosen spot. I still escort my dogs to the dog run to be certain they have eliminated. Don't expect your dog to go there on his own. One of the most important aspects of housebreaking your dog is praising him for an appropriate elimination. How can you praise the dog if you are not there with him? You must be by his side, coaxing him to go and offering accolades when he does.

When It Is Raining

Many dogs do not like to eliminate in the rain. They do not like the wetness on their feet. The substance and situation are different. Remember, dogs learn in certain contexts. Many people have housebreaking problems with their dogs when it is raining. Combat the situation by familiarizing your dog with going potty in the rain. Take advantage of every rainy day. If you take your dog out and he does not go potty, watch him closely or crate him when you bring him back in the house. Take him back outside every twenty to forty-five minutes until he goes.

Chow Time: Scheduling Food and Water

Embed this in your brain: INPUT = OUTPUT. Translation: your puppy will have to eliminate after he ingests food or water. A young puppy may have to urinate as soon as fifteen minutes or less after drinking a lot of water and will probably have to defecate right after eating solid food. By controlling when you feed your puppy, you control when he eliminates. Your puppy's food and water intake should follow a strict schedule:

- Feed the puppy at the same time every day. Young puppies should eat three meals a day: morning, noon, and early evening. Puppies three months or older are fed only twice daily: morning and early evening.
- Remove food after fifteen to twenty minutes, and do not offer it again until the next feeding. If you free-feed the dog, you will have no control over his housebreaking program, and you could also create problems with rank and training.
- Take away the puppy's water three hours prior to bed. This helps prevent nighttime accidents and allows the puppy to sleep more soundly and longer.

I generally take puppies off a three-meal program when they are approximately three to four months old. I find that the puppy is less interested in the midday meal by this time. Therefore, most puppies are going to be fed two meals per day. It is very important that these meals be fed at the same time every day because INPUT = OUTPUT. You must stick to a schedule. By feeding the dog at the same time every day, you condition his body for regular elimination.

When you free-feed a puppy—that is, have food available all the time—two problems occur. One, the puppy's bowel habits are not put on a schedule. Two, you give all your power away. This is problematic when training because you want the dog to *work* for food rather than think it's his God-given right. Also, when you free-feed your dog, he never has the chance to get truly hungry. Don't feel that you deprive your dog or puppy when you take away his food or water. If he does not eat when you begin implementing his schedule, he will soon learn that he'd better eat or else.

Stop feeding your dog by about six in the evening. This helps allow the dog to void his bowels by an acceptable time for sleep. However, you could go as late as your schedule allows. Tailor the dog's feeding time to your lifestyle. If you stay up late every night, it's okay to feed the dog later. Just make sure you take away his water three hours before bedtime. This way, both you and the dog will sleep longer and more soundly. The only exception to this is if the temperature is high and you do not have air-conditioning. In that case, offer ice chips or small amounts of water hourly.

Picking a Time

While it's really important to schedule meals and water, you need to choose a time that is convenient for you. Potty training does not mean you have to rearrange your entire calendar. Whether you wake up at five in the morning or ten in the morning, you can accustom your dog to go potty according to your schedule. Your puppy needs to adhere, for the most part, to your schedule. Please be reasonable with your potty times though. Realize that young puppies get up at the crack of dawn. It's like having a baby. You will most likely need to get up with the puppy at different times of the night and morning until he adjusts to his schedule. Also, keep in mind that this is a seven-day-a-week schedule, not one for Monday through Friday and a separate one for the weekend.

Erratic Schedules

Once the dog is older and housebroken, start to change his feeding time. That way, if you happen to be gone at a time when the dog is normally fed, he will not be completely distressed as a result. By fluctuating the feeding times, the dog can't discern when meals are served. Start slowly by varying the amount of time you are away from your dog without feeding him. I can feed my dogs anywhere from six in the morning to one in the afternoon. This is very good for people who have erratic schedules. However, if you have a very predictable schedule, I suggest you stick with the norm.

Confinement

A common error people make when housebreaking their dog is allowing the animal to wander about freely before he is properly housebroken. Confinement is the key to prevention. Your puppy should always be under strict supervision. Otherwise, he must be confined. Unless you can watch *everything* the puppy does with the eye of a hawk, you must keep him either on a leash or in a crate. Yes, it's true—on a leash, in the house. A dog will not eliminate where he lives, and until he sees your house as his living space, he will not be housebroken.

Spend time with your dog or puppy in each room of your house so that he views it as his living space. This lowers the chances of the dog soiling in uncharted territory. I was working with show-biz mogul Michael Ovitz's Golden Retriever, which was only allowed loose on the first floor. He was otherwise kept outside or in the crate. One day when the parents were gone, one of the kids took the Retriever exploring on the second floor. Sure enough, when Mr. Ovitz came home he found a big yellow stain on his white wool carpet. He was so furious that he wanted to get rid of the dog. Once he calmed down, reason prevailed and the dog won a reprieve. As far as the dog was concerned, there was nothing wrong with relieving himself in the bedroom. He did not know it to be his living space.

Use the following confinement methods until you are positive your dog is housebroken. When you cannot watch your dog, he must be in a safe location such as outside in an appropriate potty area or in the house crated. When you are with him in the house, he should be crated, on a tie-down, or under strict off-leash supervision. If you allow the dog to roam freely about the house, you invite an accident to happen. Most likely he will find a corner and relieve himself when you are not looking.

Not only do these methods help you housebreak your dog, but they help make him more comfortable on a tie-down.

Umbilical Cording

You can "umbilical cord" your dog or puppy to you by leading him on a leash as you move through the house. This is effective training for any dog since he is physically attached to you at all times. Hold the leash in your hand or attach it to a belt loop. Use a relatively short leash, otherwise the dog may be able to sneak out of your sight. A three- to four-foot leash should suffice. With this method, the dog or puppy is under your strict supervision. Because you are watching your dog or puppy does not mean his need to go potty is quelled. It merely prevents him from sneaking off somewhere and going behind your back. It is your responsibility to take him frequently to an appropriate elimination area.

Puppy Stations

A puppy station is an area of your home where you tie your dog's or puppy's leash to a piece of furniture. Make several puppy stations in areas where you tend to spend a lot of time. The den where you watch TV, the kitchen where you cook, or any other area that seems to make sense are all appropriate. Setting up a puppy station is easy, too. Attach an inexpensive three- to four-foot leash to a very strong piece of furniture. Hook the dog to the leash when you are in that room. Only use a buckle collar when you do a tie-down, and make sure to give him an appropriate chew toy to keep him occupied.

Once the dog or puppy becomes accustomed to the tie-down area, he sees it as a living space. Housebroken dogs choose to relax in the tie-down areas on their own even when they are off-leash. The point of the puppy station is that the puppy will not want to soil the area where he hangs out. He would rather go to the corner, pee, and then return to his hangout spot. If the amount of room a dog or puppy has to explore is restricted, he is much less likely to soil the area. The dog learns to associate that area as a living space, one that he would not soil. Never leave a tied dog unattended, in order to prevent him from getting tangled, or worse, hanging himself.

Crates

Crates have several purposes, but their first and foremost use is for housebreaking. Crating is the most effective way to prevent an indoor accident when you cannot strictly watch your dog. When you crate a dog, you teach him bladder and bowel control. Dogs generally do not soil their crates, an area they consider their sleeping and living space. If a dog or puppy is allowed to run free in the backyard, he will never learn how to hold it because he will be able to eliminate whenever he feels the need. A dog, like a child, must be taught the idea of bladder and bowel control. And again, as with a child, this requires time and patience. Do not think of a crate as a prison. Remember, dogs are den animals by nature. They prefer small, confined spaces, and they come to love their crates.

Taking It to the Extreme

I had two clients with a similar housebreaking problem. Refusing to crate their dogs at night, my clients allowed them to sleep in their bed. The dogs jumped off the bed in the middle of the night and eliminated at will. These clients insisted on sleeping with their dogs. Needing to find a solution, I jokingly suggested "crating" themselves in the bed with their dogs. Ironically, it worked. They put an exercise pen around the parameters of their beds, corralling themselves and their dogs inside. In the morning, they were thrilled with their clean floors, and they took the dogs outside for a proper elimination.

When to Crate

During the day, crate your dog or puppy up to three hours at any given stretch. Depending on his age, the time of day, his energy level, and your lifestyle, you can increase or decrease the amount of time he is crated. Never crate an adult dog for more than four hours, and never crate a puppy for more than two to three hours at a time. The good news is, most dogs sleep during the day; therefore, this is not the incarceration one would perceive it to be. At night, crate the dog in the bedroom with you. If the bedroom is not a possibility, then any high-traffic area like a kitchen or den will suffice. Take out your dog or puppy first thing in the morning, and immediately usher him to the potty spot where you teach him to go potty on command.

It's a good idea to exercise your dog or puppy prior to and after being crated. Exercise tires the dog, and he will want to sleep while crated. Also, make sure the dog empties himself prior to crating. Always crate an empty dog.

Dirty Puppies

A dirty puppy is a puppy that is willing to eliminate in a crate and sit in his own waste. These dogs have no concept of cleanliness. Dirty puppies usually come from bad breeders or pet stores where the dog was not given an appropriate outside elimination area and was forced to eliminate in the same place he lived. Dirty puppy syndrome is a learned behavior. Most dogs and puppies are naturally very clean animals.

Dirty puppy syndrome obviously creates a huge problem with housebreaking because you must retrain the puppy to be clean. Dirty puppies grow up without feeling that stepping or living in their own waste is incorrect. Housebreaking a dirty puppy requires twice the effort since you need to retrain the dog's thwarted instinct.

You have to teach the dog to want to be clean by keeping his living space immaculate. Shrink the amount of room the dog has in the crate. Leave him only enough room to stand up, turn around, and lie down. Take the dog out to the appropriate potty area, and when he eliminates, say "good potty."

If the dog has an accident in the crate, clean him immediately. Never allow the dog to spend any time with his mess. The dog needs to develop a new sense of cleanliness. Unfortunately, the more time he spends in a dirty crate, the more comfortable he feels with that way of life. When the dog is not in the crate, he must be with you on a leash under strict supervision or outside in an appropriate potty area. If you want him roaming free in the house, you must literally watch every move he makes.

When You Cannot Crate

If for some reason a crate is absolutely not a possibility, confining your dog or puppy to a small room such as a laundry room or a bathroom is feasible. However, some strict rules apply. Never shut the door on your dog. A closed door ostracizes and isolates your dog, creating barrier frustration. Instead, use a baby gate tall enough to restrict the puppy. Also, the room should be small enough for the puppy to view it as his living space. Consequently, he will become hesitant to soil it.

Downfalls to this method do exist. An avid chewer may nosh on your walls, moldings, and the gate itself. Also, because the dog's or puppy's movement is not restricted, he may feel he can soil one area of the room while remaining clean in another. These are strong reasons not to use a room for confinement and instead to use a crate.

Potty Times

With a puppy, potty times are as follows: first thing in the morning, after a nap, last thing at night, after chewing on a chew toy, after playing, after eating, after drinking, after everything. Young puppies need to go potty after virtually every change of behavior. Puppies between the ages of six weeks and four months are peeing and chewing machines. As they get older, the need to eliminate becomes less frequent as long as you help build bladder and bowel control by using confinement methods.

Do not give the puppy a chance to make a mistake. Take him out every twenty to thirty minutes to one hour, depending on the age of the puppy and the time of day. Note that dogs are most active at dusk and dawn. Therefore, dogs tend to eliminate more frequently during those times. They will naturally be more awake between six and ten in the morning and five and nine in the evening. Be prepared for this schedule, and understand that it is normal.

Many clients often wonder why their dogs can sleep through the night without an accident but can't last for six to eight hours during the day without having to go. This is a ludicrous expectation for our dogs. A dog's metabolism changes depending on his activities. At night, the metabolism shuts down as it does with us. There is little stress and stimulation on the body. In the daytime, the metabolism is up and functioning, and thus excretions are much more frequent. Also, during the day the dog ingests food and water, and since INPUT = OUTPUT, the dog will have to go potty more often.

Potty on Command

You can easily teach your dog to eliminate on command. As you take your dog outside, say to him, "Outside. We are going outside." Outside simply means you are going to the great outdoors. It does not necessarily have anything to do with going potty. It is merely a location. As soon as your dog or puppy begins to eliminate in your chosen potty area, start chanting your potty command. Mine for lack of anything more original is "potty," but you may use any command you choose, such as "get busy." Your dog's potty command should be used for both potty functions. I find "potty" most efficient.

As the dog eliminates, which means the pee or the poop is falling from his body, start to quietly chant, "Potty, potty, good potty, potty, good potty, potty." Chant your potty mantra in a low, subdued tone because if you sound too excited, you might distract the dog. Initially, it should almost be said under your breath.

If the potty chant is done correctly, it becomes a command that elicits a conditioned response. It's like when you turn on the faucet and the running water makes you feel like you have to pee. Later, once the dog knows the command, you say "go potty," and even though there is a bird, a leaf on the ground, or some other distraction, the dog thinks, "Yeah, I do have to go potty." It's as if the light bulb in his head turns on. Potty on command is extremely helpful and time efficient.

Treats

Potty training is the one area in which I find treat training to be distracting rather than helpful. Many dogs become too preoccupied with receiving a treat. A dog that always gets a treat when he eliminates may learn to urinate an amount small enough to get his reward. The owner, thinking the deed is done, brings the dog into the house, where the dog empties the rest of his bladder.

However, if once in a blue moon I am outside with my dog chanting a potty command and I happen to have a treat, I will surprise her with it. At the very moment the dog finishes eliminating, I hand her the treat and say "good potty." With a dog that has never experienced going potty outside, I occasionally use a treat to reinforce and encourage this new behavior. Let's say you have got a holdout, which is a dog that only pees in the house, in the crate, on the carpet, or anywhere but outside. He has never eliminated outdoors. The first time he has a breakthrough, give him a treat for hitting ground zero.

Will He Tell You He Needs to Go?

You can teach your dog to tell you he has to go outside, but it is better for you to be more aware of his elimination habits. Beat him to the punch. If you still want a way for your dog to let you know that he needs to go, always say "outside" before you take him there. Any potty cues your dog might give you, such as standing or barking by the door, acting antsy, or circling and sniffing, should be met with a chorus of "Outside. Do you have to go outside?" Once the dog has made the connection between the potty cues and "outside," he will learn to tell you when he needs to go. Be wary of the dog that catches on to this and uses it to his own advantage. He may sometimes tell you he has to go outside when he really doesn't have to go potty. He wants to go out just for the fun of it and views you as a revolving door.

Accidents

An accident is a big deal because it means *you* are not doing your job. It is your responsibility to teach your dog what is a proper or improper elimination area. Don't expect him to know and understand what you want from him right away. After all, we give children a substantial amount of time to be potty trained; we should at least afford our dogs the same courtesy. Just because your dog goes outside once does not necessarily mean the behavior is set in stone. Keep a close tab on your dog until you are positive he is 100 percent housebroken.

How and When to Deal with an Accident

Be instructive when dealing with accidents. If you catch your dog or puppy in the act of an inappropriate urination or defecation, DO NOT say "no." DO NOT say "bad dog." Instead give the instructive command of "outside" in an even tone of voice, pick up the dog or puppy, and usher him to the appropriate elimination area. When you pick up the dog, he usually stops midstream, allowing you to get him outside. Make sure not to use a harsh tone of voice, because you may frighten the dog. A frightened dog does not learn that going in the house is wrong; rather, he learns that going in front of you is wrong. If you are lucky and the dog continues to go in the appropriate elimination area, praise him by saying "good potty, good potty outside."

There is absolutely, positively no punishment after the fact. If you walk into a room and the dog has already eliminated, you are too late for recourse. Whatever you do, never push your dog's nose in his mess or hit him. This is cruel and it is dog abuse. If you reprimand or punish your dog after the deed is done, he will have no idea what he has done wrong, and you will end up scaring him. The only thing you can do when a mistake is made is clean the area with an odor neutralizer and get right back to a strict housebreaking regime.

Cleanup

Don't clean mistakes in front of your dog. While cleaning the mess off your carpet, you will most likely have a bad attitude, which your dog may sense. This could result in a dog that is scared to eliminate in front of you. Put your dog or puppy in another room, and use a good odor neutralizer to remove the scent. Once a dog anoints any given area, he is habitually attracted to that area. Mistakes are less about scent and more about habit. All the odor neutralizer in the world may not stop your dog from returning to an area he has soiled once before.

Health Check

Sometimes a well-trained dog or puppy suddenly makes a mistake. If the problem persists, a health check may be in order. Sudden inappropriate elimination may be a sign that your dog has a physical ailment or is stressed. Diarrhea may be a sign that your dog has parasites or may be the result of a dietary change. A dog or puppy that urinates frequently may have a urinary-tract infection. If he urinates small amounts in inappropriate areas, take him to the vet for a urine analysis. A urinary-tract infection is usually cured after about a week or two on medication. Note that some medications such as prednisone, a common allergy medication, increase urination, so don't be surprised if you have a lapse in housebreaking. Whenever your pet is put on medication, ask your veterinarian what behavioral or physiological side effects you might observe in your dog.

Stay on Top of It

If an accident occurs, it is on one hand a major setback and on the other no big deal. Once a dog has soiled an area, Pandora's box opens. The dog's mind thinks, "Why bother going outside when I can go right here?" Do not give your dog the opportunity for a mistake to happen again. You can quickly eradicate an inappropriate elimination by returning to a strict housebreaking program.

Reverse-Housebroken Dogs

A reverse-housebroken dog is a dog that wants and has been conditioned to eliminate in the house. This situation can be corrected by forcing the dog to go potty outside. Pick a time when you are prepared to be with your dog for a two-day stretch.

Crate the dog at night and take him outside first thing in the morning to the lovely grass area you have chosen. The dog will probably walk around a bit and then look at you as if to say, "I don't do grass." Bring the dog back into the house, place him in the crate, and feed him food and water. You want to increase the pressure on the bowel and bladder so that when you next take the dog outside, he feels he is about to burst. If the dog continues to withhold, repeat the procedure until he finally lets loose in the new elimination area.

Once the dog has eliminated in the new elimination area, it's as if a taboo has been broken. You must understand that the dog is not trying to be stubborn. He has simply never experienced going potty in this foreign area. Once the dog eliminates in the appropriate potty spot, the next time will surely be easier, followed by the next and the next and the next.

Paper Training

Paper training is not a viable training method. Be wary of a breeder who brags that his puppies are paper trained. Paper training absolutely contradicts the goal of teaching the dog to go potty outside the house. When you train a puppy on papers, he learns to go potty in the house. The papers become secondary. If you take away the papers, the puppy usually continues to eliminate in the same area.

A dog that is paper trained is a dog that is not truly housebroken. He is absolutely happy to go inside the confines of a house. This is why most paper-trained dogs have accidents. Whether there is paper or not, the dog has learned to eliminate in his living space, thereby failing to respect the space as a place to be clean. A paper-trained dog may go for a walk and wait until he returns to the house to eliminate.

Occasionally, I have clients who do indeed want their dogs to go exclusively on papers. They are generally people who live in high-rises or are in other ways prohibited from taking the dog outside for a proper pee. These people are in the minority. Even when living in a high-rise, if one has a balcony, a potty box is preferred to papers. A potty box is a two-by-four-foot box of sod. Not only does the dog learn to go potty in an outside elimination area, but he also learns to go on grass. A potty box gives the high-rise person the best of both worlds.

Such was the case with actor Mike Myers and his yellow Labrador Retriever puppy, Borchevski. Mike and his wife, Robin, lived in a high-rise. For those evening and early morning toilet trips, it was easier and more timely for them to use a potty box on their balcony. During the day, they continued to take Borchevski outside. Eventually, as the puppy grew up, they phased out the use of the potty box.

No Way Out

If paper training is your only option, then follow these steps. Do not leave the dog on the papers unattended. Treat the situation as if you are going outside. Confine the potty spot to a small, uncarpeted area such as a bathroom or laundry room. Use an appropriately sized baby gate to enclose the area. Cover the entire surface with papers. Butcher's paper is preferable because it is sturdier than newspaper, and because it does not have ink, which can bleed.

When your dog begins to eliminate, start the potty chant, and when he is finished, immediately remove him from the potty area. Your dog should naturally gravitate to one area of the papers. Once your dog's potty spot preference becomes clear, begin to shrink the amount of space available to him until he is left with a two-by-two-foot area, the approximate size of an opened newspaper.

Dog Doors

Dog doors are the lazy person's way of housebreaking a dog. Perhaps the dog learns to go outside, but in my experience, the moment the dog door is closed, the dog pees right in front of it because he has never experienced having to hold it. He is a dog that has not truly learned to be housebroken. A dog door is permissible once the dog is properly housebroken.

A dog cannot be 95 percent housebroken. He either absolutely abhors eliminating in his living space or he has carte blanche to go there. If I were somehow prohibited from getting home to my dogs, days later they would be found in my house dying to go outside. Lotte and Ruby are truly housebroken dogs. They would rather burst than eliminate in the house. This is not because they think peeing in the house is wrong, but because it's simply not a possibility for them. Going outside is the only thing they have ever known, and it is therefore the only possibility in their little dog minds.

Walks

There are two different types of walks: elimination walks and exercise walks. An elimination walk is a short trip outside with the sole purpose of allowing your dog to void himself. An exercise walk is a longer walk taken for fun. When you take your dog for an exercise walk, he should eliminate prior to the walk, either in his potty spot or at the very onset of the walk. If you confuse the two types of walks, the dog quickly learns that his fun, exercise walk ends as soon as he eliminates. What happens then is you walk and walk and walk, and the dog holds out to prolong the walk.

If you do not have a yard with an appropriate potty spot, and your dog's only means to eliminate is on a walk, then take him on elimination walks. The only purpose for this type of walk is elimination. If the dog knows the potty command, then use it when you take him out. Otherwise, teach him to go potty on command on these short walks. If the dog or puppy does not eliminate when you take him out, then bring him back into the house, but watch him very closely. After twenty minutes or so, take him back outside. Continue this process until your dog finally eliminates. You might want to reward your dog's hasty potty on an elimination walk with an exercise walk. Eventually he will connect a hasty potty with a fun walk and no potty with a return to the house.

Marking

Do not let your dog mark. Keep walking. For one thing, he does not void all his ammunition in one place and instead learns to hold it and dole it out in small increments. Next, marking is an aggressive behavior. An intact male that marks as he walks down the street is saying, this lamppost is mine, this tree is mine, this ditch is mine. An intact male acts as if he owns the block when he marks. This behavior can be highly problematic with a dog that has dog-aggression problems. Also, it takes twice as long to walk a short distance because the dog keeps stopping to mark.

Submissive and Excitement Urination

Submissive and excitement urination is not a housebreaking problem, but rather a matter of temperament. When a puppy greets an older dog, he often turns on his back or stands upright and salutes with a small amount of urine. What this behavior tells the older dog is, "Hi, I'm a puppy. Please don't hurt me." That is exactly what the puppy says to us when he greets us with a dribble of urine. This behavior is more common in female dogs. Most dogs grow out of this by around the age of one. However, I know of several dogs that exhibit this lovely behavior their entire lives. This is also a breed-specific behavior. The poster dog for submissive or excitement peeing would be the American Cocker Spaniel. They simply cannot help themselves.

Never punish your dog for this type of urination. Submissive and excitement peeing are absolutely involuntary behaviors, and therefore not within the dog's control. The behavior occurs on a physiological and emotional plane. It can only be corrected through positive reinforcement and desensitization to people entering your home and greeting your dog. If you try to reprimand the dog, you will complicate matters and make the situation much worse.

The worst way to try to correct this annoying behavior is to call attention to it. Instead, keep your arrivals low-key. Do not make eye contact with, touch, look at, or give attention to the dog until he gets over the novelty of your homecoming. When approaching an extremely submissive dog, try walking toward him sideways in a crouched position. By not approaching him head-on, you take on a less dominant posture. Ask your guests to practice low-key arrivals until your dog quells the behavior.

The line between submissive and excitement urination is blurred. Generally, the submissive pee-er crouches to the ground with his ears lowered back, and the excitement pee-er wags his tail and dances in his urine while he pees. However, if the dog has been perpetually reprimanded, you can be sure the behavior is submissive urination rather than the dog being excited to see you.

Control from Chaos

A few years ago, I was called to work with Barbra Streisand's four-month-old Bichon Frise, Sammy. In the month and a half that she had the dog, many housebreaking problems occurred. Sammy was kept loose in the kitchen, where the staff unsuccessfully attempted to paper train him. When they took him outside, he would not go potty and would return inside to soil.

I immediately brought a crate into Sammy's life. His food and water were strictly scheduled, and he was crated at various times throughout the day. I put Sammy on a strict elimination schedule and made a chart on which all staff members could note when Sammy eliminated and whether he urinated or defecated. The schedule enabled them to make certain that Sammy eliminated before they took him to Barbra in other areas of the house. Sammy is a beautiful illustration of how order can be made out of chaos.

In Short

Housebreaking boils down to two things: preventing the dog from having an accident and praising him for appropriate elimination. If you can religiously get your dog to an outdoor elimination area, he will view the inside of your home as his living space and will work hard to keep it clean.

Dogs are creatures of habit, so be prepared. Before your dog or puppy comes home, pick an appropriate elimination spot.

Schedule all meals and water because INPUT = OUTPUT. Take away water three hours before bed.

Confine the dog when you cannot watch him. During the day, crate a puppy up to two to three hours and an adult dog up to four hours. At night keep dogs crated in the bedroom, laundry room, kitchen, etc.

At all other times, the dog should be with you on a leash, under strict off-leash supervision, or with you on a tie-down.

Potty times for puppies are:

- after a nap
- first thing in the morning
- last thing at night
- after playing
- after chewing on a chew toy
- after drinking or eating
- after everything

Do not give a puppy a chance to make a mistake. Take him out every twenty to thirty minutes to one hour, depending on the puppy's age.

To teach elimination on command:

1. Say "outside!" as you go outside.
2. Take the dog outside to your chosen potty area.
3. As the puppy eliminates, say "good potty."

When there is an accident:

- If the puppy is caught in the act, say "outside" and hurry the puppy to the appropriate elimination area.
- NEVER punish after the fact.
- Remove the puppy from the area, and clean up the spot with an odor neutralizer.

Paper training and dog doors are crutches that usually make housebreaking more, rather than less, difficult because the dog never builds up bladder and bowel control. The dog is not reliable in the way that most owners desire.

There are times when a dog just cannot help himself and soils inappropriately. The dog may be ill or may be responding in a submissive or excited manner to an overwhelming stimulus. Take the dog for medical treatment for the former, and work on confidence building and desensitizing the dog to whatever triggers the urination for the latter.

10

The Mechanics of Training

Although we come from an entirely different perspective than our dogs, we do have the ability to communicate with them. If we expect to live with dogs, we must integrate them into our lives by teaching them our language. It would be unfair for us not to train them. Our job as parents, owners, and teachers of dogs is to train them and make a connection with them. It is our duty to help our dogs understand our world, since it is our world in which we are asking them to live.

Why Train?

Training your dog is more than teaching him tricks to impress your friends. It is the most humane act you can do for your dog. A trained dog is less likely to get into trouble or enter a dangerous situation. A trained dog is absolutely free, whereas an untrained dog is segregated outside, kept in a crate all day, or eventually given away because he didn't make the grade. An untrained dog never learns to be assimilated into our lives.

A trained dog is a free dog because he understands how to live within our world. Being able to allow your dog more freedom takes a lot of pressure off you. A trained dog understands a "stay" or "no street" command, and will be less likely to run into the street and risk being hit by a car when you open the front door or car door. You can leave him in the house without fear of him destroying your belongings. Training also helps with poison control since you can teach the dog not to pick up random items from the ground. If he does find something to nibble on, you can say "off," and he knows to spit it out.

Training prevents your dog from doing anything you do not want him to do. When you give your dog a command, there are a million other things he cannot do because he is busy obeying you. If you give your dog a "down-stay" command, he can't dig in the yard or jump on the couch because he is too occupied concentrating on the command.

Communication

One of the most important reasons to train your dog is communication. When you teach your dog commands, you teach him our language and you begin to interact with him. He better understands what you want, and you may even begin to understand what your dog wants. Think about the people in your life. Those with whom you communicate well are probably your strongest relationships. It's the same with your dog. Open lines of communication yield a healthier, more reciprocal relationship.

Your dog's vocabulary really can become quite sophisticated. When it does, you can breathe a little easier, trusting that you can give your dog a command and that he will understand and follow through with it. Establishing good communication with your dog takes quite a bit of patience. Think about when you first meet someone who does not speak English. The two of you use signs, sounds—anything you can—until you come to an understanding. It can be extremely frustrating, but when you do finally get across to one another, you are elated.

So Much to Say

One night while soaking in my Jacuzzi, I had a slight communication problem with Ruby, which was easily corrected by Lotte. Keep in mind that at the time of the incident, Ruby was still an adolescent, and no matter how much training a dog has, when he is young, he will make mistakes. You also need to consider the dog's breed. Remember that certain breeds are more independent than others and it is their nature to do their own thing.

In the Jacuzzi was a white, spongy ball the size of a tennis ball that floats around collecting dirt. Appropriately, it is called a scumball. Ruby was completely taken with it. I dunked under the warm water, reemerging to find Ruby and the scumball gone. From the Jacuzzi, I could see Ruby in my bedroom on top of my bed chewing on the scumball. Of course this would not do. It was totally inappropriate behavior on her part.

I called an "off" command to her from the Jacuzzi. Ruby backed away from the ball. I called "off" again, and she hopped off the bed. Great. One problem was solved, but another still existed. The dirty scumball was still on my nice, dry bedspread. Feeling a little lazy and very comfortable myself, I did not want to get out of the Jacuzzi to retrieve a scumball. Not wanting it on my bed either, I called Lotte into the situation.

I told Lotte, "Lotte, in the house." She walked from the Jacuzzi deck through the open French doors into my bedroom. Then I said, "Lotte, dog toy, find the dog toy." She zeroed in on the scumball, but in her excitement, she dropped it. "Take it," I encouraged her. When she had it in her mouth again, I told her to "come," and she did. When she was back outside, sitting in front of me with the scumball in her mouth, I held out my hand and said, "give." Lotte put it right in my hand.

Considering her age, Ruby showed a high level of obedience. Many dogs would not have left the ball, but would have tried to play keep-away. Ruby showed good comprehension. It is not something that happens overnight. Some dogs will never reach the level of understanding that Lotte has.

Masters of Body Language

Dogs are masters of body language. They communicate everything—dominance, submission, elation, playfulness, fear, aggression—through body postures. They often learn hand signals before they learn verbal cues. People are always impressed when their dog follows a command from a hand signal alone. However, learning hand signals is a fairly simple task for your dog. The real stars of puppy class are the dogs that follow a verbal command without the crutch of the hand signal.

How You Sound and Look

You need to use a tone of voice that reflects the meaning of the command. An "off" command should sound more serious, whereas a "come" command should sound more welcoming and lighthearted. In the beginning, exaggerate your tone to fit the command, like bad acting. You may need to sound like a psycho growling "Off!" and the next second sweetly saying "take it" for a correct response. Initially, your dog responds to your tone of voice. Once your dog begins to associate your tone with the meaning of the word, you can shift to more normal tones. Eventually you want to be able to simply whisper commands.

As for body language . . . you want a confident, imposing stature. This means you need to face your dog with a tall posture and squared shoulders. A lot of people incorrectly bend low to the dog's level when they give a command. Your body needs to face the dog straight on. If you need to bend, don't crouch but bend a little at the waist. You will have more authority. The only exception is the beginning stages of the "come" command because crouching to the ground can make you seem more welcoming to the dog.

Leash on Life

The leash should be thought of as an extension of your arm. It enables you to communicate in what direction you want your dog to go. The leash also gives you control over your dog and helps you guide the dog through training. There is not one standard way to hold a leash. Find what feels comfortable to you. I usually hold the leash in my left hand in order to free my right hand for hand signals. Generally, I hold the leash low, with my hand naturally hanging by my side. Depending on the exercise, I tighten or loosen the slack of the leash. For example, when walking I loop the leash around my hand to shorten the lead, but for "stay" I will lengthen it so that I can move away from the dog. Because a leash gives you more control over your dog, the more you use one, the more your dog becomes accustomed to following your commands. Strengthening your dog's compliance on a leash brings you closer to having strong off-leash commands.

Pay Attention

When you start training your dog, you need to stay on top of things. Don't give him a command and then casually turn to your friend for a chat. Your casual posture signals to the dog that the work is done. While you're bragging to your friend about how smart and well trained your dog is, he's up and running. If you want your dog's attention, you have to give him yours. It is simply not fair to expect your dog to stay focused when you are not.

By continually making eye contact with your dog and by maintaining a strong posture, you command attention from him. It is also imperative to always follow through with your commands. When you give your dog a command, make certain he obeys. Otherwise, your dog learns that you may or may not be serious about commands.

Double-Commanding

It helps to begin training your dog with double commands. Double-commanding is the simultaneous use of a hand signal and a verbal cue. An untrained dog has no idea what you are saying. "Sit, sit, sit" probably sounds like "blah, blah, blah." You don't know what "blah" means, so why should your dog? You must start with something your dog can understand. A visual cue is a fantastic springboard. When you train your dog by double-commanding him, he learns to associate the hand signal with the verbal command.

The beauty of double-commanding is that you end up with two

effective ways to communicate with your dog. Your ultimate goal is to get a response from a verbal cue only. Without the visual lure of your hand, the dog must engage his total thought process. It is a higher form of communication.

What Every Dog Should Know

Regardless of which command you are teaching your dog, certain principles apply:

1. Get the action started by luring the dog with a treat.
2. Name the action.
3. Reward and verbally praise the dog specifically regarding the behavior

Some trainers believe you should only give a command once. I believe it is better to be more helpful. By giving positive feedback in the form of treats and verbal praise, you solidify the dog's understanding of the command. Give your dog a verbal and visual reminder every now and then, and reinforce the behavior with feedback. Also, when you verbally praise your dog, your word choice needs to be specific to the behavior. When your dog sits, don't say "good dog"; say "good sit." It is another opportunity to name and reinforce the behavior.

If your dog does not follow your command, do not reprimand him unless he absolutely understands what you are asking of him. A dog cannot do what he does not understand. You can encourage your dog with a tug on his leash. The key word is *encourage*. I say this with much trepidation. Do not jerk on the leash or drag your dog on the leash. If needed, use the leash to guide your dog in the direction you want him to go.

When you start training, do it in the house or a private area where there are few distractions. Keep the training light and fun. It should never be a chore for your dog. You should only train for five to fifteen minutes several times a day, depending on your dog's age and attention span. Frequent, short periods of time keep training interesting. Quit after the dog gives his best effort. Once you've had a breakthrough, stop. Resist the inclination to try to get the dog to do it again. Always end on a high note.

Teaching Your Dog His Name

In order to train your dog, you must have his focus and attention. One of the best ways to get him to focus on you is by teaching him his name. The name signals the dog to pay attention, telling him to look at you because you are most likely going to give a command. You could also use other phrases like "watch me" or "look." These can all be taught quickly with the help of a few treats. To teach a dog his name:

1. Show the dog a treat.
2. Raise the treat to your eye area while saying your dog's name. Your dog's line of vision should follow the treat to your face.
3. Oscillate the treat between the dog's eye and your eye.
4. When the dog looks at you, say his name, tell him "good," and give him a treat.
5. Next, allow the dog to become distracted, call his name, and reward him when he looks at you.
6. Repeat the process many times until your dog knows his name.

Bring the treat to your eye so that your dog becomes accustomed to looking at you when he hears his name. You know you have your dog's attention when he makes eye contact with you. At first, use an upbeat tone when you say your dog's name, and never use your dog's name as a negative. Many clients imply their dog has done something wrong by adding a tone of disappointment when they say the dog's name. Do not do this. Instead of screaming the dog's name when he jumps up on your dinner table, tell him "Off!" By using instructive reprimands, you give him an appropriate alternative rather than only admonishing him for something he thinks is normal behavior.

You don't want your dog to think his name means something bad. When he is a puppy, use his name for calling attention to something positive. Once a dog is correctly trained, understanding both his name and instructive reprimands, he will not associate his name with a negative. If you fall into the trap of implying disappointment via his name, he will understand. Don't stress out. You have not ruined your relationship with your dog. Also, in a multi-dog house, unless you call your dogs by name, how else are they supposed to know which is the culprit?

Sit

When you have your dog's attention, you can teach him to sit. Sit is one of the easiest commands to teach your dog:

1. Hold the treat between your index finger and thumb with your palm up.

2. Hold the treat between the dog's nose and mouth, so the dog can smell and even lick at the treat.

3. Tell the dog "sit" as you raise your hand slowly over his head, allowing him to follow the treat with his nose.

4. As the dog raises his head, his rear should lower to the ground. *As soon as his rear touches the ground,* tell him "good sit" and give him the treat.

The movement of your hand physically lures the dog's body into a sit. At first, keep your hand near your dog's nose and mouth. Possibly even let him lick the treat. If your hand is too high, he will jump up in an attempt to reach the treat. Once the dog is proficient at that level, stand straight and raise your hand, bending at the elbow. Next, begin to use the hand signal and verbal cue without the treat in your hand. Gradually change the treat from a lure to a reward. The final stage is using only a verbal cue.

Sit command with proper palm-up hand signal; treat and hand raised above dog's head

It is imperative to reinforce and reward a sit, or any command, at the exact moment it happens. If your dog does not sit after several attempts, you can physically help him into position, but as with most training, it is much better to use a hands-off method. Otherwise, you may always need to push your dog into position. Try combining the use of food as a lure with physical manipulation of the dog's body.

Bring the treat over the dog's head while you tuck his butt under. Again, as soon as his rear touches the ground give him the treat and say "good sit." If that fails, hold the dog's collar, and place your other hand on his rear or hind legs. Push down the dog's rear while gently pulling up his collar. This method should only be used on the rare dog that is absolutely disinterested in treats or toys. The dog will often comply after the first or second physical manipulation. Return to the treat-only method as soon as possible.

For down command, palm is down with the treat between thumb and index finger.

Dog follows the treat to the ground.

Once dog is down, the treat is given as reward.

Down

Down is sometimes difficult to achieve because it is a vulnerable position for your dog. Your dog may be a bit reluctant to go down at first. Be patient. To teach your dog to lie down:

1. Start with the dog in sit. He is already halfway there.
2. Hold the treat between your thumb and index finger with your palm down.
3. Let the dog try to nibble the treat but keep it somewhat concealed between your fingers.
4. Never allowing the dog's mouth to stray from the treat, slowly lower your hand to the ground between the dog's front paws while you say "down." When his nose and the treat are on the ground, draw the treat away from the dog. You will be making an L-shaped movement. Your hand moves down and slightly away from the dog.
5. As soon as the dog's body hits the ground tell him, "good down" and give him the treat.

The dog should follow your hand down to the ground in an attempt to free the treat. Once he learns the command, your hand signals become more subtle, and you get to the point where a verbal command alone is sufficient to get the dog down. If the treat and your hand are not strong enough lures, you may have to physically move the dog down.

With one hand gently place pressure behind the dog's shoulder blades and lure the dog to the ground with a treat in your other hand. You can also carefully pick up one of the dog's front paws while lightly pressing your other hand behind the opposite shoulder blade. By lifting the dog's paw, you create a tripod effect. Take the dog down to the ground by gently pushing on his shoulder blade and pulling slightly on his leg. Be careful not to push straight down, but rather bring the dog down on the same side as the lifted paw. Immediately reward the dog with a treat, and praise him with "good down." After a couple repetitions using a physical prompt, wean the dog off the need to be placed into position. Return to the hands-free method.

Physically place dog in down position by lifting paw to create tripod effect.

Push dog down and to the side by pulling his leg and pressing his shoulder.

Stay

Stay is a bit trickier than other commands because your dog will naturally want to follow you. No matter what distractions surround your dog, he should remain in whatever position you have asked him to assume. To teach stay:

A stop-sign hand and front-facing body posture are needed to teach "stay."

1. Put the dog in sit or down.
2. Stand directly in front of the dog, tell him "stay" and put your hand in front of his face like a stop sign: palm flat and forward.
3. Give the dog a treat every few moments that he sits, and praise him saying "good stay." Continue to remind the dog to stay between accolades.
4. Keeping your face and body directed at the dog, begin to step around him. Continue to reward the dog while he stays.
5. Release the dog by saying "okay" in a cheerful tone, but do not give him a treat.

When you start teaching stay, your dog should be on a leash. A leash gives you more control in case the dog breaks from the stay. Keep your treats in the same hand with which you hold the leash. Periodically retrieve a treat with your stay hand, give it to the dog, and quickly return to the stop-sign hand signal. Many people hold the treat in the same hand that they use for the stay command. This is an awkward position. It draws the dog or puppy to jump up and take the treat from your hand. A flat palm directly in the dog's face inhibits him from getting up.

If at any time during the exercise, your dog breaks from the stay, lure him right back into sit or down. Stand directly in front of him, and hold your stop-sign hand in his face. Say "stay" and give him treats. If you are casual about stay, your dog will be casual about it, too. Give him your full attention, and do not give him the treat if he breaks from his stay.

When to Give the Treat

The most important part of teaching your dog to stay is to reward him during the behavior. Many people make the mistake of telling the dog to stay, releasing him, and then rewarding him. Accustom yourself to the stay ritual. Tell the dog "stay," then say "good stay" and give him a treat. "Stay," "good stay," give him a treat. "Stay," "good stay," give him a treat. Constantly reinforce and reward the dog during the behavior, whether you are standing in front of him or walking around him. You can very quickly increase the amount of time between verbal reinforcement and treat rewards. When you do release him, do not give him a treat, or he will associate the reward with the release and not the stay.

Walking Around Your Dog

When walking around your dog, make sure you continually face him. Do not turn to your side and walk a circle around the dog. Slowly take baby steps to the side. Attempt to sidestep a semicircle or horseshoe around your dog. At first you may only be able to take a step or two before your dog gets up. The longer your dog stays, the farther you should try to walk around the dog. Try to make it halfway around one side and back to center and then halfway around the other side and back. Your goal is to walk a full circle around the dog. If the dog is in a down, you will actually get to the point where you may be able to step over his back. When you start to go behind the dog, he may get nervous, so you want to gradually work to that level.

Keep a stop-sign hand and a front-facing body posture while walking around the dog in a "down-stay."

Keep Him Down

Initially your dog may only give you a couple seconds of stay. That's okay. Build on it. Reward those two seconds while the dog complies. You develop a strong stay very slowly. If he wiggles a lot, place your hand behind his shoulder blades or withers while you tell him "stay," put the stop-sign hand in his face, and give him a treat. Obviously, in this case,

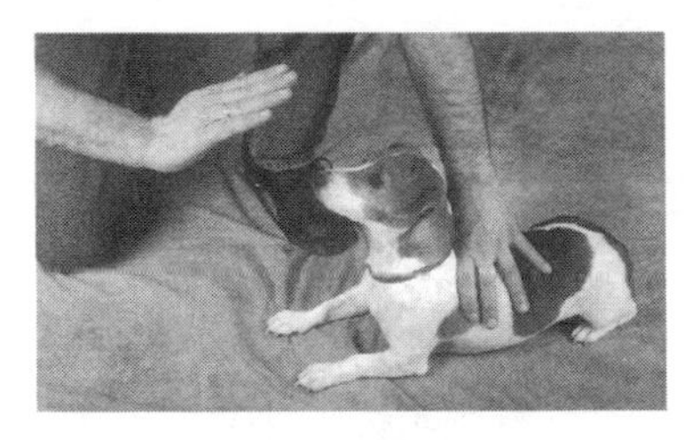

Control a squirming dog by holding him down with one hand while using your other as a stop sign.

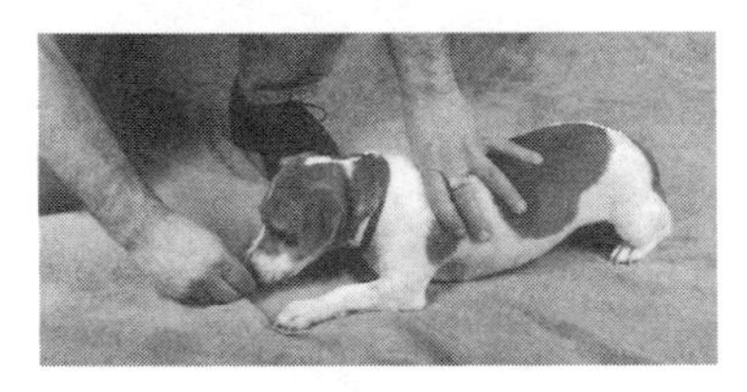

Continue to hold a squirming dog while rewarding with a treat.

you won't have a free hand to hold your reserve of treats, so keep them in your pocket, where you can easily reach them.

Always keep one hand on a squirming dog. Use the other hand to make the stop sign and to give treats. Give the dog the treat, and boom, put that stop hand right back in his face. Do not give him a chance or reason to get up. When the dog stops wiggling, use the hands-free method. The next stage is to begin to step around him in a semicircle. Progress until your dog perfects his stay.

Distance, Distraction, and Duration

The best way to proof your dog's stay is to increase competing motivation. Start by adding distance. Reinforce by stepping in to give him his treat. Next, add distractions. You may need to employ help for this one; kids work well, and bouncing balls are a great distraction for most dogs, too. Finally, add duration to the equation. Increase the length of a down-stay to fifteen to twenty minutes and a sit-stay to no more than three minutes. Constantly reinforce and give feedback, particularly with stay. This is a difficult command because it requires much concentration and willpower from the dog.

Lower your criteria for your dog, and add each competing factor separately. If you are working on distraction, get really close to the dog while the balls are bouncing around him. Shove your stop-sign hand in his face and treats in his mouth. Don't even worry about distance or duration at that point. It's the same idea with duration. Focus only on how long the dog stays, and continually reinforce a good stay. When working on distance only, concentrate on how far you can walk away before your dog moves. Remember to face the dog even as you step back.

The goal is to combine all three: distance, distraction, and duration. Try for a five-minute down-stay while you are twenty feet away and balls are flying past your dog's head. Add each one separately, and try different combinations as the dog's concentration improves. You gradually work your way to a three-ring circus.

Come

The most common complaint with regard to obedience training is probably "My dog won't come when I call him." Come is one of the most important commands that we can train our dogs to respond to. It is also one of the hardest commands. We expect our dogs to come to us no matter what the distraction, obstacle, or competing motivation. He is supposed to stop chasing a squirrel or tear himself from any other interesting activity and come to us. This takes a large amount of conditioning as well as an extremely strong leader-follower relationship. To teach come:

1. Show your dog a treat and let him sniff it.
2. Walk backward with the leash in one hand while holding the visible treat toward him in the other hand, and tell him "come" in a happy tone.
3. Reward him with the treat and say "good come" when he comes to you.
4. Add a "sit" command once the dog comes to you.

To teach "come," walk backward while luring the dog with a treat.

The come command must equal come and sit. Otherwise, when you are off-leash in the park and you call your dog to come, he may very well trot toward you and continue past. In his mind he fulfilled the command by running to you. It won't even occur to him to stop unless he knows "come" means he also has to sit when he gets to you. It is also a good idea to touch his collar just before you give him a treat. Should you need to grab him in an emergency, your dog will already be familiarized with you reaching for his collar.

The advanced hand signal for come is a wave of your hand toward your chest. As your dog's training progresses, make come more difficult by initiating it from a stationary position like a sit-stay or by letting the dog get distracted before you call him to come. You will need to increase the length of your leash when you call him to come from a distance.

Walking away from the dog while teaching him to come is integral because dogs are attracted to movement. Lure him with a treat, your movement, and a goofy, fun attitude. Coming to you should be more fun than anything else around him. Never do anything the dog may perceive as negative when you call him to come, such as:

- reprimand him; "come here you bad dog" is a cardinal sin
- clip his nails
- bathe him
- make him go into the house
- crate him

Reserve the come command for something special. You can put any of these negative actions on command. Take advantage of your dog's potential for vocabulary, and name those other actions. Tell him "go to your crate," "inside," or "bath time."

Non-come-pliance

Use your leash to encourage a dog that doesn't come, but don't drag him with it. Give an encouraging tug. The leash should only be used to get the dog moving. Once he is moving, the treat in your hand, your backward movement, and a fun tone of voice should be substantial enough to draw the dog toward you. If there is no competing motivation, if the dog likes your treat, if you have substantially and positively reinforced the behavior in the past, and your dog still does not come when called, then you need to examine the situation. Either you have done something incorrectly, or your dog does not understand the command. If noncompliance continues, there may be a more serious problem; your dog may not view you as his leader. Brush up on your leader-follower relationship. Also, make sure come is always a pleasant command, successfully rewarded and reinforced.

Off

"Off" means remove your paws, your body, or your mouth from whatever you are touching or thinking about touching. When given the off command, your dog is to get off the person he's jumping on, get off the couch, stop chewing the wicker chair, or spit out the chicken bone he found. To teach off:

1. Conceal a treat in your hand so that your dog can smell it but can't really see it or get to it.
2. Hold it about eight to ten inches from the dog's nose.
3. Tell him "off" in a low, meaningful, no-nonsense tone.
4. When he moves away, even slightly, tell him "good off," reveal the treat, and tell him "take it."

The key to successfully teaching the off command is rewarding the dog that ceases his attempts to take the treat. When he displays an avoidance gesture by turning his head from the object of his desire, the treat, he is rewarded. An avoidance gesture is a dog's way of saying he is no longer thinking about what is interesting to him. He totally defers to the owner, his leader, when he turns his head and shows a lack of interest.

Teaching "off" with a treat visible in an open palm

After your dog begins to understand when you say "off" with the treat hidden in your hand, progress to the next stage. Tantalize him by exposing the treat in the palm of your hand. He may try really hard to get the treat, but don't let him. Be stern with your "off," lowering your voice. If needed, you could try to embellish it with the stomp of your foot or guide the dog away from the treat using the leash; with a young puppy, try a muzzle grab. Once the dog passes the open-palm stage, increase the level of difficulty. Teach him the off command with the treat on the floor and then with it between his paws. Eventually, you will be able to put him in a down-stay, place treats on his paws, tell him "off," and be confident that he will not move.

When you place the treat on the floor, it becomes open season. It's much easier for a quick dog to grab the treat before you do. DO NOT allow the dog to get the treat. When the treat is on the floor or his paws, call attention to it by pointing at it. If he goes for the treat, your hand will already be positioned between his mouth and the treat, and you can immediately block him. Again, growl "off," and use a noise embellishment. A muzzle grab may be appropriate for a puppy. When the dog turns away or backs off, your immediate response must be "take it." The duration of time he abstains from grabbing the treat gradually lengthens.

A treat on the floor is open season. Be quick and stay close.

Take It

You need to teach your dog to take the treat just as you need to teach him to stay off it. He will not know the meaning of the words until you teach it to him. When you tell him "take it," hand the treat to his mouth. You probably won't have to struggle to get your dog to eat a treat. Never let your dog pick up the treat on his own. Pick it up off the floor or his paws, and hand it to him with permission to take it. You don't want your dog to get accustomed to taking things off the ground. "Off" and "take it" can act as poison control. "Take it" also develops into a retrieving command. It tells the dog to put an object in his mouth.

A Confrontational Command

In order to teach the off command, you must tempt your dog with something he finds appealing, like a treat. Essentially you tell the dog, "I know this is something you want very much, and even though I am holding it in your face, you cannot have it." Saying "off" in a serious tone of voice while staring the dog in the eye is confrontational.

Be very careful in your assessment of the dog's temperament. The off command should be taught to puppies under five months of age or to extremely docile dogs. Some dogs will think the direct confrontation of you saying "off" is unacceptable, and they will come up against you. If you are at all frightened by your dog's temperament and think he may retaliate in an aggressive way, do not attempt to teach him this command. Seek competent, professional help.

Walking

While walking, your dog should be at an even pace with you. Traditionally, dogs are trained to walk on the left side. It does not matter which side you choose; pick one and stick to that side. Consistency is important.

The traditional use of the word *heel* sounds militaristic, nonsensical, and demeaning. I use two commands: "let's go" means move forward, and "easy" means slow down. Both of these commands are much more conversational.

Dog is at your side and on a leash when learning walking commands. Maintain position by luring with a treat.

Let's Go

To teach your dog to move forward:

1. Walk with the dog at your side.
2. Lure him along with a treat, and say "let's go."
3. Reward him with a treat for keeping up with you.
4. Every time you stop, tell him "sit" and raise a treat over his head.

Easy

To teach your dog to slow down:

1. Walk with the dog at your side (again, I will use the left side).
2. Tell him "easy," and hold the treat back by your side. If he pulls, say "easy," then give a slight leash correction.
3. Again, reward him for maintaining the proper pace and position.
4. Every time you stop, tell him "sit," and raise a treat over his head.

Start by using a buckle collar, but if your dog pulls excessively, you may consider using one of the many devices available to curb this behavior while teaching the easy command. A no-pull harness or head halter helps relieve the pain in your arm and the pressure on your dog's neck. In a rare case, you may use a pinch collar with professional guidance. Make sure you hold the leash low, in a natural position. It should also have enough slack so that you are not pulling on the dog, but not so much that he can walk ahead of you. Your goal is to keep him right by your side.

Break It Down

It's helpful, convenient, and more communicative to use two commands for walking. "Let's go" tells the dog to move in the same direction and with the same pace as the owner. "Easy" tells the dog to slow down and keep pace. "Come" is not a viable walking command because it tells the dog to go from point A to point B and sit. It's more fair to break the commands into specific increments rather than having a general command for several actions.

Street Protection

Street protection is actually boundary training. The same principle applies to teaching your dog not to go in the pool, through the front door, out the gate, or in a certain room of the house. A "no street" command sets up a boundary. Your dog will learn not to run into the street after a ball, another dog, or anything else. I use "no street" for lack of imagination. You can get as creative as you like. To teach your dog not to run into the street:

1. Stand at the curb with your dog. Tell him "sit."
2. Step in the street without the dog, and say "no street." Give him a treat.
3. Step back up on the curb, say "good no street," and give him a treat.
4. Slowly increase difficulty for your dog by jumping and then running into the street while he stays on the curb.
5. Immediately say "good no street," and give him a treat for not following you.

When teaching street protection, don't let your dog follow you into the street. If he starts to step off the curb with you, guide him back onto the sidewalk with the leash, and tell him "no street" in a low tone of voice. Once he is back on the sidewalk, tell him "good no street" and start the process again. Practice this whenever you get to a curb while walking your dog. One day while walking him, he will stop and sit when the two of you reach the curb without you giving him the command. Your dog must never venture off the sidewalk unless accompanied by a person and given permission. Tell him "okay, let's go," and the two of you can then cross the street together.

Pool Protocol

Bridget Fonda's German Shepherd, Sunflower, discovered the fun of swimming. Bridget did not mind Sunflower splashing around in the pool until her swimming habit turned excessive. She would swim when no one was around to dry her off. A wet dog means puddles, soggy rugs, and messy floors.

Bridget and I taught Sunflower a "no pool" command. I stood between Sunflower and the pool and said "no pool." If she moved toward it, I strengthened my tone of voice or stomped my foot as a sound embellishment. When she backed away from the pool in the slightest, I rewarded her with treats and praised her by saying "good no pool."

Sunflower was a bit older and was already trained. She understood "off" and "no street," which made "no pool" extremely easy to teach. After one session, she had it down. I could throw balls into the pool and tell her "no pool," and she would not budge. No matter what we did, Sunflower would not dive in without permission. Amazingly, she stayed out of the pool even when home alone.

Distance Commands

Teach distance commands to get your dog to do something from a distance. This is beneficial in emergency situations. If you are twenty feet from your dog and you tell him "sit," you want him to stop and sit where he is. A dog that does not understand distance commands will ignore you or come to you and then sit. Your dog must have a strong understanding of the down and sit commands before he can learn distance commands. To teach distance commands:

1. Tie your dog's leash to something stationary, like a doorknob.
2. Walk away.
3. Tell him "sit" or "down," and use the corresponding hand signal.
4. When he complies, praise him with "good sit" or "good down," and *step in* to hand him the treat.
5. Walk away again.

Distance commands are more difficult because the dog has learned the sit or down command in a certain context: directly in front of you. Double-command to help your dog; he has a better chance of understanding you when you use both a verbal and a visual cue. You must also step in and give your dog his treat. If you've got good aim and your dog is a good catcher, you can toss him the treat. Either way, your dog needs to receive the reward in the place that he is doing the behavior. If you let the dog come to you to get the treat, you reward him for coming, which is contradictory to a distance command.

Emergency Sit Distance Command

Distance commands can be used in numerous emergency situations. Because come is such a difficult command, the chances of getting a dog to sit are much higher even when you are a great distance away. Sit is usually the first command a dog learns and the one used most often. The sit command is generally a dog's strongest trained behavior. An emergency sit distance command can circumvent many dangerous behaviors and situations.

Go to Your Place

It's a great idea to give your dog his own place in the house. You can then train him to go there on command. You can get specific with your locations by teaching "go to your bed," "go to your crate," or "go to your place." It's nice to have an area in different rooms of your home that your dog may come to know as his own. A practical use of this command is when you're eating. You can send your dog to his place without banishing him from the room. To do this:

1. Show the dog a treat and let him sniff it.
2. Lure him to his place, say "go to your place," and give him the treat.
3. Next, toss a treat to his place, and say "go to your place."
4. When he follows after the treat, tell him "good place."
5. After a few repetitions, say "go to your place," and point toward his place.

6. When the dog goes to his place, tell him "good place," and step in with or toss him a treat.

You need to step in to give your dog the treat because he needs to be rewarded when and where the behavior happens. Once you can send him to his place by pointing, make the exercise more difficult by adding a down-stay command once he gets there. Soon, "go to your place" will mean "go to your place and down-stay."

Have Patience

Environmental visual cues help you recall what you've learned. Contextual learning is stronger for dogs than it is for humans. Our brains are much more developed and, therefore, adapt better to different situations. Dog learning is very literal. Each command has a specific body language and reference point.

We need to revel in the beauty of the possibility of communicating with another species. We can talk to dogs, but it requires patience and it takes hundreds of repetitions of a command. Most people expect their dogs to understand them from the start. When our dogs don't understand us, we punish them. Would you punish someone who didn't understand your language, or would you try to teach it to him or her? We have higher standards for our dogs than we do for other people. It's wondrous to communicate with your dog. Once you accept your dog as a different species, you will stop taking the ability to communicate with him for granted.

In Short

Training is the key to your dog's freedom. One of the most important reasons to train your dog is the rich communication you will share.

Like dogs, we communicate through vocalizations and body postures. When giving your dog a command, use your voice effectively by enunciating the command, growling the instructive reprimand, and warming up the tone when issuing praise. Keep your body posture tall, and engage in eye contact with the dog. In the beginning, double-command your dog by using both a hand signal and a verbal cue.

Once you have captured your dog's attention, each command is taught in a similar manner:

1. Start the action by luring the dog with a treat.
2. As the dog's body is moving into the new position, name the action—e.g., sit.
3. Reward and verbally praise the dog while he is doing the commanded behavior; e.g., say "good sit" while the dog is sitting.

When teaching a new behavior, never reprimand your dog. That should only happen when you are absolutely sure that he knows the command and has chosen not to comply. When in doubt, take a step back in your training to where he was last successful, and move forward from there.

A well-trained pet dog will respond to the following commands: his name, sit, down, stay, come, off, take it, let's go, easy, no street, and go to your place. Each is initially taught in a comfortable, low-distraction environment. As the dog masters the command, you can increase the difficulty by varying distance, distraction, and duration. Vary only one element per exercise, and increase the difficulty in small steps to enable your dog's success while learning control in increasingly more difficult environments.

We can communicate with our dogs, but it requires patience and hard work. However, the benefits for both parties make it all worthwhile.

11

Behavior "Problems"

There is no bad behavior to a dog. However, what is absolutely normal for dogs may not neatly fit into our lifestyles. Consequently, we brand them as bad or aggressive dogs. There is always a reason and a purpose behind any dog behavior that has been instilled in their makeup for generations. The bottom line is, they are a different species than us.

When a client comes to me with a "problem," I don't immediately spew out advice. The dog's temperament, breed, age, and health need to be considered. I then require information surrounding the behavior. The first step to solving any behavioral problem is finding out why the dog is doing that particular behavior. What are the circumstances surrounding the behavior? When does the dog do it? Where does the dog do it? How does the dog do it? It is also usually a good idea to have a medical checkup to rule out the possibility of a physical problem. After all this information is considered, I devise a program to change the behavior.

Exercise and Mental Stimulation

Many behavioral problems occur when a dog is bored. One of the best ways to combat boredom is to make sure your dog is well exercised and mentally stimulated by giving him plenty of opportunities to interact with his environment and other beings. Many people share the misconception that a big backyard and a cozy house is enough to keep a dog entertained. If you were forced to stay home day in and day out, you would probably begin to feel cabin fever. It is no different for your dog.

Dogs, like humans, need an abundance of stimuli in their daily lives. Unfortunately, many dogs only leave their property for an occasional walk. This is not enough exercise, nor does it give the dog a chance to experience the world. Even some zoos are now realizing the importance of mental stimulation for animals. They are incorporating environmental enrichment programs to make the animals' daily lives more interesting.

Although a backyard may have a variety of stimuli, it is incorrect to think that all your dog's needs will be satisfied there. A backyard is only as good as who or what is in it with the dog. Your dog must be regularly exposed to a variety of new sights and sounds. Play, training, walking, hiking, car rides, and dog parks are all great forms of exercise and fabulous opportunities for interaction, thus keeping your dog from becoming bored. Keeping your dog mentally stimulated and well exercised can help prevent inappropriate behavior.

Separation Anxiety

Let's talk about your favorite loafers that you love and wear all the time. You've spent a long day at work, and you come home to find that Buttercup, your Rottweiler-Shepherd mix, has decided to munch on the left loafer. You sigh and take it to the closet, only to find she's peed all over the right shoe. Buttercup's trying to tell you something, right? Wrong! This is one of the biggest misconceptions people have about dogs. When they destroy your property, they are not mad at you or trying to get back at you. They may, however, be experiencing some sort of emotional distress, which their behavior reflects. They may be suffering from separation anxiety.

Oftentimes when people bring home a new dog or puppy, they overbond with him. They shower him with affection and coddle him to the point where he won't leave their side. Then, when Monday comes and they have to go to work, the animal is terrified because suddenly he is alone. He has lost his pack. The dog is not angry. He is anxious and scared to death that his owners are never going to return. His destructive behavior is a reflection of his intense separation anxiety.

The specific manifestations that the separation reflex can take are:

- jumping through windows to get out of the house to find you
- chewing and scratching at the door through which you left, trying to get at you
- howling or barking in an attempt to call you back
- loss of bowel or bladder control from intense anxiety
- tearing up and chewing at your bed, pillows, clothes, or personal belongings

Overbonding with a Rescue Dog

Separation anxiety is even stronger with rescue dogs. Because they may have been through a traumatic experience, rescue dogs have a tendency to be very needy. As a result they readily bond and the new owners feel so much guilt over what the dog has potentially been through that they overcompensate by coddling the dog. They want the dog to feel safe and to feel like theirs is the last home he will ever have. The combination of the dog's need to feel safe and the new owner's sense of guilt can result in overbonding. Much to the owner's naïve delight, the dog follows her from room to room.

Loving Him Back to Health

Courtney Thorne-Smith got a one-and-a-half-year-old Basenji named Ed from a Basenji rescue organization. Courtney's story was no different from anyone else's. She gave Ed a period of adjustment during which she wanted to love him back to health. Ed was soiling the house, chewing, and becoming extremely anxious when Courtney left. We applied a classic separation-anxiety treatment program, which worked beautifully.

To treat separation anxiety:

- Establish rules and boundaries immediately.
- Teach the dog to sit or lie down for anything he wants. Don't shower him with free affection.
- Provide plenty of exercise and mental stimulation.
- Possibly confine the dog to a crate or other small area with comforting items like his bed, chew toys, etc. Frequently enter and exit the room until he begins to calm.
- Come and go many times from the house. During the exercise, ignore the dog for the first few minutes you are there, and then leave again, gradually lengthening the amount of time you are gone.
- Practice low-key departures and arrivals. Don't make contact with the dog for ten to fifteen minutes before you leave and after you return.
- Assume a strong leadership role. Dogs truly love a strong leader.
- Never punish after the fact. You will increase the dog's anxiety.

Please note: In severe cases, consult a veterinarian, who may prescribe antianxiety or antidepressant drugs for your dog.

How to Avoid a Nightmare Situation

To avoid a separation-anxiety nightmare, leave the dog right off the bat. Instead of focusing all your attention on the dog, do just the opposite. Show the dog that he can survive on his own. It is actually counterproductive to give the dog a period of adjustment during which you want to let him know he is loved. You will love your dog right out of the home because you have set him up for failure.

A Matter of Taste?

Back to those loafers. Why does the dog always pick your favorite item to destroy? Does the dog have good taste? No. The dog seeks out objects that carry a lot of your scent. Generally, the things you love maintain your scent because you tend to handle them or wear them much more than your other items. Buttercup has most likely chewed your loafers because you wear them much more than your five-dollar specials. It is a way for the dog to envelop you and feel close to you in an attempt to alleviate some of his anxiety. He is not being vindictive or feeling guilty. Dogs do not think that way. He is relieving stress in a way that is available to him. It is like our habits of wringing our hands or pacing when we are upset.

Chewing

Dogs need to chew. Chewing is a normal function that keeps their jaws, teeth, and muscles in proper condition for eating a carnivorous diet. Even though domesticated dogs no longer have the need to kill their prey, they still demand an outlet for chewing. Problems occur when people give untrained or young dogs access to many inappropriate objects. Always look at the bigger picture. Your dog may be chewing because he is anxious, bored, or merely underexercised.

Inappropriate Chew Toys

Dogs acquire preferences for certain materials; therefore, do not give your dog that which resemble household objects to use as chew toys. For example, if you give a dog a rope chew toy, he may develop a preference for the feel of cloth between his teeth and may readily chew the fringe of your Oriental carpet. Also avoid fleecy toys, stuffed animals, socks, or anything made of cloth. A cloth chew toy can be a gateway to your rug or sofa. However, not every dog will develop this habit.

There are other toys that are not only much more appropriate but also will be preferred by the dog. Because dogs acquire a preference for substances, you can intervene by giving your dog a proper chew toy made of an appropriate substance, thereby creating a proper chew-toy habit. You decide what you want your dog to be accustomed to chewing.

Play Toys

Many people argue that their dog loves his stuffed toy. It's his "bunky bear." This may be fine if the dog merely plays with a stuffed toy. There is a difference between chew toys and play toys. Stuffed toys, squeaky toys, and tennis balls are fine for the dog to play with. If he is chewing or dissecting the toy, then he may be developing a habit that could lead to the destruction of some of your coveted possessions. Also, the eating of play toys is not safe for the dog.

I had one client with a yellow Labrador Retriever named Sophie. He gave her plastic water bottles to play with. He ranted and raved about how much Sophie loved to chew water bottles. One day, Sophie jumped up on the kitchen counter and grabbed the plastic drain mat from under the dish rack. In the process she knocked over the drying rack full of crystal vases. Sophie chewed the rim of the mat and, for days, pooped plastic.

Appropriate Chew Toys

Use chew toys made of material that is unlike anything else you have in your house. Some appropriate chew toys are:

- sterile bones—real cow bones that you stuff with dog treats
- raw cow bones—knuckle or femur are best; cooking causes them to splinter
- Kong toys, which look like little snowmen made of rubber; hollow inside and can be stuffed with a myriad of food items
- flip chips—large, round pieces of rawhide shaped like a corn tortilla; avoid rectangular ones

- pressed rawhide—rings, cartoon bones or dumbbell; avoid particle rawhide

Pig ears are more of a food item than a chew toy. Dogs eat them rather quickly, and a chew toy with more longevity is preferable. I am also not too keen on my dog ingesting a shellacked pig's ear.

Flip Chips

One of my favorite chew toys for young puppies is round flip chips. Choose ones that are bigger than the dog's mouth. I love them because puppies are unable to swallow them and instead chew them on their back molars. Because of their large, round shape, flip chips render themselves safer than the popular rectangular-shaped ones, which are more readily available. My motto with chew toys is, the bigger the better. As with a child, the smaller the toy, the greater the danger of swallowing and possibly choking. When and if a chew toy becomes too small, throw it away.

Rawhide

The grand pooh-bah of chew toys is pressed rawhide, sometimes called Buffalo Bones. Do not confuse this with particle rawhide, which is chipped rawhide that is pressed into another form much in the same manner as particleboard. Pressed-rawhide chew toys are sheets of rawhide compressed into a shape. They come in five-, eight-, ten-, and twelve-inch lengths. Choose a size that is larger than your dog's mouth.

Unfortunately, pressed-rawhide chew toys are not always easy to find at the local pet store, but they can be acquired through catalogues. Buy natural, tan-colored rawhides made in the United States. The white ones are often bleached, and foreign brands are often cured with harmful chemicals.

Don't get these smooth, pressed-rawhide chew toys confused with the kind of rawhide chew toys with knots on their ends. Those can unravel, becoming slim and easily swallowed by the dog. I have pulled out long pieces of rawhide from a dog's throat. The flimsy, knotted rawhide and the particle-rawhide chew toys break apart very easily and are quickly ingested. A properly chewed pressed-rawhide chew toy is systematically pulled apart bit by bit, rendering the small pieces harmless in the dog's stomach. However, some dogs may have difficulty digesting even small bits of rawhide, resulting in vomiting or gas.

Nylon bones are shaped like cartoon bones and are made of hard plastic. They are supposedly really safe. I think they are safe because few dogs show interest in them. They are too hard to chew for many dogs and, therefore, uninteresting. Somewhere down the line a vet, breeder, or friend put the fear of rawhide into dog owners who opt for these other boring chew toys. Consequently, the dog's need to chew is not satiated and he turns to household items.

Pay Attention

Whenever your dog puts anything in his mouth there exists a potential risk. There is always a possibility the dog will choke or ingest an improper amount. As with children, we need to monitor our dogs' activities. Since chewing is a dog's natural behavior, we cannot and should not avoid it but should use prudence and care during chew time. It is unfair to deny a dog the chance to chew.

Often an active or powerful dog like a Labrador Retriever or Pit Bull will eat an entire rawhide chew toy in under half an hour. This usually occurs when the dog chews the wrong type of chew toy. That is too much rawhide to be in a dog's stomach at any given time. You may want to try a larger, twelve-inch pressed-rawhide chew toy because it will take longer for the dog to consume. Monitor the dog while he chews, and take the chew toy away after twenty minutes. If you are giving your dog the proper rawhide chew toy and you still find it disappears too quickly, it may be that your dog cannot have a rawhide chew toy. You may opt for stuffed sterile bones, stuffed Kong toys, or large, raw, cow knuckle or femur bones instead.

Keep It Interesting

When I look around houses with balls, nylon bones, fringe toys, and stuffed animals, I see a house without chew toys. I then reach into my

Mary Poppins bag and pull out a proper chew toy. Not only do you need to have the proper chew toys, but you also need to keep your dog interested in them. Otherwise, other objects, such as the leg of your coffee table, may become more interesting to him. Some tips:

- *Soften rawhide chew toys.*
  Sometimes the rawhide bone is too hard for your dog. You can soak the chew toy in a cup of hot water or hot chicken or beef broth. Marinate it for five to ten minutes. You can also use a spritz of a spray butter product or other flavor enhancer.

- *Flood the dog with chew toys.*
  Dogs that have been rawhide-deprived may become too chew-toy obsessed. If you flood them with chew toys, their overinterest will fade to a regular level of interest.

- *Rotate chew toys.*
  Keep chew toys in a basket away from the dog, not on the floor. Rotate three to four different toys. This does not mean substituting a rawhide for a sterile bone, but rather having four different toys down at one time and then substituting those with four new versions of the same toys.

- *Praise the dog while he chews.*
  It's a good idea to name chew toys and praise the dog for chewing appropriate chew toys. When you praise him, you help create a positive reinforcer. Common phrases I use are "good chew toy" and "good dog toy."

- *Stuff chew toys with treats.*
  You can stuff Kong toys and sterile bones with any treat the dog loves: freeze-dried liver, dog jerky, peanut butter, cheese, whatever. The dog will be rapturous trying to get out his hidden surprise.

Prevention

Like most behaviors, chewing is best solved by prevention. Do not allow your dog access to anything you do not want him to chew. When you are

A crated puppy with a stuffed sterile bone happily develops an appropriate chew-toy habit.

gone, crate or place the dog in a safe area. When you are home, keep him in the same room with you on a tie-down or directly supervised with appropriate chew toys. The goal is for the dog to never be in a position to learn that chewing the couch, carpet, or woodwork is fun. You will reach a point where you trust your dog to roam freely in the house. I crated Ruby until she was two and a half years old because I did not trust her not to chew in my house. Now she is left loose in the house with zero destruction.

In the event that your dog does chew inappropriately, you can only reprimand him during the behavior. Do not come home and yell at him while he is sleeping next to a shredded pillow. While his mouth is on the inappropriate object, tell him "off." If that fails, stomp your foot or use an antichew spray, and tell him "off" once more.

Barking

There are many reasons why dogs bark: they bark for attention, because they want something, because they hear a noise, out of fear, out of aggression—for as many possible motives as there are emotions. You need to assess why your dog is barking by examining the circumstances in which he barks. Keep in mind, some breeds bark more than others. Once you conclude why your dog is barking, you can remedy the situation.

What's He Saying?

You need to discriminate between different types of barks. Alarm barks, attention barks, and fear barks occur for diverse reasons and sound different. Most problems occur with recreational barking, which happens when the dog is left alone in the backyard or house. He barks because he is bored and unhappy and is trying to summon his pack—you!

A dog that barks at you is a dog that is laughing in your face or demanding something from you such as food or attention. This is much different from a dog that barks because of separation anxiety. You treat the demanding dog by restructuring the leader-follower relationship. The dog must learn to view you as the leader so that he does not think he can tell you what to do. Sometimes that means using a reprimand. Of course a fearful dog that barks from separation anxiety can sometimes be made more anxious by the use of a negative. Likewise, the use of a negative may cause an overly confident dog to become aggressive with you. Care must really be taken when using a reprimand.

Eliminating the Problem

Barking when you are at home is much easier to remedy because you are there to interrupt the behavior. Teach your dog the quiet command, or do something to show your grave displeasure with his behavior. If the quiet command alone does not work, try it in conjunction with a sound deterrent such as the noise made by stomping your foot or smacking the wall. If the barking persists, tell him "quiet," and employ a spritz of breath spray in the dog's mouth. You may try a muzzle grab with a young puppy.

As with any reprimand, keep your dog's personality in mind. When he does not bark, remember to reward him for being quiet, but choose your moments. If your dog does not bark at a time when he would normally—for example, when looking at a squirrel—say "good quiet" and possibly give him a treat.

Always make sure your dog's needs are being met. If he's left alone too long, he may need attention, and if he's bored, he may need more exercise. If your dog barks because he hears a noise outside or is distracted by a squirrel or the neighbor's cat, then bring the dog inside. The problem can sometimes be fixed that easily. If bringing the dog inside is not a possibility, then get a squirrel guard or put up a barrier so that the dog can't see the object at which he is barking. Do what you can to get rid of the source. If there are a multitude of distractions and the barking persists, then use a bark collar.

Bark Collars

Bark collars are not cruel. However, it is cruel to get rid of a dog because his incessant barking makes your or your neighbors' lives miserable. It is also cruel to continually yell at and punish a dog that barks. When other attempts to quiet your dog fail, using a bark collar is not an outrageous tactic. It is more humane to use a bark collar on your dog than it is to give him up or, worse, put him to sleep simply because he barks.

Citronella Collars

Citronella collars are an extremely useful and humane tool to stop unnecessary recreational or inappropriate barking. They use a sensor that detects the vibration of the dog's throat when he barks. They then emit a spray of citronella under the dog's chin. Citronella is harmless and has been veterinarian approved.

Electronic Bark Collars

If your dog is unfazed by citronella spray, you can try an electronic bark collar *as a last resort.* Instead of a spray, these collars emit a static shock. With a good quality collar (like Tri-Tronics new Bark Limiter) on a low-level setting, the sensation for some dogs is similar to the feeling when you hit your funny bone. I recommend collars with at least five to seven levels of stimulation. You need to find the appropriate level at which your dog learns to be quiet without hurting him. Most dogs learn to curb their overall barking behavior with these tools or products. Other dogs become collar-wise. They only comply when they wear the collar. Electronic bark collars may be unacceptable to some people; however, I have found them to be humane and effective in the few cases where they are needed. Please consult a professional trainer or behaviorist when considering using this collar. Tri-Tronics can be reached at 800-456-4343.

Your Dog Will Not Become Mute

Do not worry—your dog will not become mute if you suppress inappropriate barking. If anything, your dog will become a much more discerning and discriminating barker. Dogs are genetically predisposed to bark, some breeds more than others. Never worry that eliminating inappropriate, unwanted barking will somehow prevent the dog from barking at the necessary time.

Barking on Command, a.k.a. "Speak"

Teaching your dog to bark on command may be helpful in teaching your dog when not to bark. You create an on-and-off switch for barking. When your dog naturally barks at something—for example, the doorbell—tell him "good speak" and give him a treat. Have someone ring the doorbell again, and tell your dog "speak." When he barks, tell him "good speak" again and give him a treat. After several repetitions, you will be able to say "speak" without the aid of the doorbell and your dog will speak. Reward him appropriately.

You may use any word you like to name the behavior. I taught Lotte to speak by saying "defend." For many people this is a very useful command. When you walk down the street and you see a questionable character or someone starts to hassle you, you can tell your dog "defend" and he will bark. The menacing person may not realize that "defend" is simply a command to bark and may think instead that you have an attack dog.

Jumping

When a dog jumps on you, he is probably trying to perform the normal canine greeting of face licking. One of the easiest ways to deal with this is to meet the dog's needs by coming down to his level. Establish that you only pet the dog when all four of his paws are on the ground. Low-key arrivals and greetings are another way to handle jumping. By maintaining a stoic composure when greeting your dog, you lessen his excitement and explosive behavior.

Old Wives' Tales About Jumping

We've all been told how we should handle a jumping dog. As usual, many of these procedures may be ineffective and even cruel. Large dogs usually do not respond to the following tactics. Smaller dogs may be too sensitive and delicate for these methods. You may injure or possibly even maim the dog. DO NOT use the following methods to stop your dog from jumping:

- *DO NOT knee the dog in the chest.*
  Not only is this inhumane, but it can be very dangerous. By kneeing the dog in the rib cage, you can inflict great harm. Also, some dogs couldn't care less. To a large and powerful dog, this could feel like a love pat.

- *DO NOT step on the dog's toes.*
  This takes a great amount of dance prowess, including choreography and agility. If you have heavy feet, you can crush a dog's toes. A large, bouncy dog may not respond at all.

- *DO NOT hold on to or squeeze the dog's front paws.*
  This may cause the dog to panic and attempt to bite you.

- *DO NOT say "down."*
  This is a training command that means "lie down."

More of What You Should Do

Teach the dog not to jump by conditioning him to sit whenever you or anyone comes into your home. When a person walks through the door, tell your dog to sit and give him a treat while he's being greeted. Continue this process every time someone makes an entrance into your home. Begin to vary the amounts and times when treats are given. Your dog will learn to automatically sit when you let someone into your home.

Teaching your dog the off command is also very useful. Do not say "down" or "no" when your dog jumps. Use the instructive reprimand of "off," which tells the dog to remove his paws from whatever he is touching. If this does not have enough power, you can back it up with something unpleasant like body blocking, that is, moving your body into or toward *their* space. Another option is to give the dog a spritz of breath spray into his mouth accompanied with the instructive reprimand *off*. This is called pairing a negative. Later, saying "off" will come to represent a negative experience, and the dog will put all four paws back on the ground. Also, if you put your dog on a leash and step on the end of it, you will have more control over his movement. He should have enough lead to sit or stand comfortably but not enough to allow him to jump up. Having your foot on the leash prevents your dog from jumping up, causing him to correct himself.

Be warned that the cute puppy you allow to jump up on you may grow into a jumping dog. It is unfair to accept this behavior when he is a puppy and then expect the dog to stop the behavior as an adult. If saying "off" does not work, a muzzle grab can be effective on a young, nonaggressive puppy. The best way to manage jumping is to countercommand the dog to sit.

Digging

Digging, too, is a normal dog behavior. Sometimes dogs dig to make a cool, denlike space for themselves in the summer. Some breeds, like Terriers, are more prone to digging than others. Sometimes dogs are after prey like gophers. Usually a digging dog is bored and needs to release some energy. It's also fun for a dog to dig. The best way to counter digging is to prevent it. If you've already got a dog that digs, then make sure he is well exercised and mentally stimulated.

Addressing the Issue

Again, prevention is the key. Drain your dog's desire to dig. An exercised dog is usually too tired to dig. If you have to leave him in the yard all day, take him for a long walk or run before you leave for work in the morning, and give him appropriate chew toys. If your dog is occupied with other matters, he will not feel the need to dig.

If you have a small space in which you do not want your dog to dig, such as a potted plant, sometimes putting rocks over the area or filling it with gravel will do. You can also camouflage chicken wire by staking it to the ground and covering it with dirt. Digging on a grand scale can be more difficult to combat. Remember, no matter what, there is absolutely no punishment after the fact. This makes digging somewhat problematic to handle since it usually happens when the dog is left alone. However, if you catch the dog in the act, use the off command. Unfortunately, the off command may not stop the dog from digging while you are gone.

Fences

Set up boundaries around your yard to prevent the dog from digging. You can put them around the borders of your yard or garden. Put up an inexpensive garden fence that you can buy at any home or garden shop. These consist of white or green wire, look like a miniature picket fence, and come stacked in a bundle. They are cheap and easy to install.

If the problem persists, there is a new citronella containment system called Virtual Fence that consists of a collar worn by the dog and a wire buried along the area that you want to keep the dog away from. When the dog approaches the area, a warning sound goes off. If he continues toward the area, a harmless spray of citronella is emitted under his chin from the collar. This may seem harsh to some of you. Let me assure you that it is not mean. Using a citronella collar is much more humane than hitting your dog or giving him away because he tore up your tulips. The dog quickly learns not to go in those areas of the yard.

Digging Pits

If you have the room, you can build a digging pit. A digging pit is a large, specified area filled with sand. Normally, that's all you need to attract the dog to dig. If not, bury a bone or other tasty chew toy to attract the dog to the area. The problem with digging pits is that many people don't have the space to spare. A digging pit is a good idea if you have a Terrier, which is a natural-born digger. You have to drain his digging instinct somehow, and it is better to give him an appropriate outlet rather than to try to cap it.

The Good News

Not all dogs dig, and most dogs grow out of their digging desire. It is usually young- or adolescent-dog behavior. My dogs only dig when I take them to the beach. Sand is a digging aphrodisiac. Dogs also prefer freshly churned dirt because it is more sandlike, more fun for the dog, and therefore, more attractive to him. This is one reason why you do not want your dog to watch you garden and why you should not let him in the yard right after you have planted.

Running Away

A dog that runs away tends to see the grass as greener on the other side. These dogs are usually bored, neglected dogs that live entirely in the backyard, isolated and ostracized from the family. He could also be a dog that does not leave the property often. When given the opportunity, he bolts out the open door seeking adventure. The other kind of dog that runs is the intact male who roams the neighborhood searching for females in heat. You can remedy a running away problem by:

- bringing the dog into the house
- showing strong leadership qualities

- giving the dog obedience training
- making the dog feel he is indeed a family member
- defining the dog's rightful place in the pack
- neutering
- boundary training
- taking the dog for walks
- simply not giving the dog the opportunity to run

Don't give your dog an opportunity to get away. A dog that gets out must be stopped. Even a dog that is happy with his family may choose to take off if he is bored. Secure your property 100 percent. Take extra care if you have a Houdini on your hands, and do whatever it takes to ensure that your dog does not get out. A low fence must be made higher. If he's crawling underneath it, secure it with cement. Whatever the problem, fix it. Period. Don't complain that your dog pries open the gate when it is your responsibility to put a better latch on it.

Other Crimes and Misdemeanors

Prevention is usually the best way to combat a dog that digs in the trash, sits on the furniture, or steals food from the counter. Put the trash out of the dog's reach, or don't leave out food. You can cover furniture with clear, plastic carpet runner turned upside down so the grip side faces up.

There are also booby traps you can use to catch a dog in the act of any of these crimes. Motion-sensitive noise devices sound alarms when the dog moves them. A modified mousetrap, specifically made for dogs, flips in the air when touched. The difference between this trap and a true mousetrap is the modified version does not snap on the dog or harm him. Electric mats come in sizes suitable for furniture or strips perfect for countertops. When the dog touches the mat, he gets a mild static shock. The mats are also useful for boundary training a dog. Place them in doorways to prevent the dog from entering a room. All these products can be purchased at the pet store or through catalogues. There is another great citronella product called the Spray Barrier. This is used indoors and has a spray collar and transmitter dish. Place the dish in an area that you want your dog to stay away from or off of, such as countertops, furniture, litter boxes, or trash bins. Similar to the Virtual Fence, as the dog approaches the area it receives a warning sound and, if the dog continues, a harmless spray of citronella will be emitted under the dog's chin from the collar. Many of these products can be purchased at pet stores or through catalogues. ABS, the company that makes the citronella products, can be reached at 800-627-9447.

Pica

Pica is the eating of inanimate objects like rocks, safety pins, pennies, anything. It is fairly rare and potentially dangerous. One theory is that pica occurs because of boredom. Another is that there is a dietary imbalance. Ask your vet if you should make any feeding changes. Obviously prevention is the best solution. Unfortunately, you cannot always control what your dog finds on the ground. This is one reason the off command is so important. It is the quickest way to get your dog to take his mouth off any object. Another solution is exercise and mental stimulation. A tired dog is a good dog.

Coprophagia

More commonly known as poop eating, coprophagia is absolutely normal dog behavior. Dogs will dine on their own poop, that of other dogs, or cat poop, the ultimate canine delicacy. Cat fecal matter is high in fat and high in taste; many dogs love it. Sometimes coprophagia is health related. If your dog begins to eat feces, the first step is to get him a health check. The behavior could be a sign of pancreatitis, intestinal infection, food allergies, or a mineral deficiency. If the veterinarian's report is clean, then you may simply have a dog that likes to eat poop.

Coprophagia is generally thought to be done out of boredom. Dogs left alone for extended periods of time usually indulge in their own fecal matter. A bored, impressionable puppy may have nothing better to do than to develop an interest in his own poop. Once he has developed the interest, it becomes a habit. To break the habit, don't give the puppy access to the feces. As soon as your dog or puppy eliminates, pick it up and get rid of it. Keep the space immaculate. Put the cat box where your dog cannot reach it, and keep the litter clean. If it is not available, they cannot eat it.

Snack Wagon

Bridget Fonda's ten-month-old German Shepherd, Buckethead, likes poop. He began his habit by eating his own poop and then moved on to the poop of Sunflower, Bridget's three-year-old German Shepherd. Buckethead followed Sunflower around like she was a mobile snack wagon. As do most owners, Bridget abhorred this behavior. Never before had she owned a dog that engaged in coprophagia.

Her best cure was to circumvent the behavior. I advised Bridget to go out with Buckethead each time he eliminated. The second he finished, Bridget picked up the feces and disposed of it. After two months of religiously picking up after the dogs, Bridget stopped and the behavior returned.

I suggested that Bridget make immediate cleanup after Buckethead a lifelong habit. It not only prevents the dog from eating his own waste material but is more hygienic and helps prevent dirty puppy syndrome. A dog that is comfortable being around fecal matter may become accustomed to walking, sleeping, and living in it. Not only is cleaning up after your dog in public a courtesy, but in most places, it is the law. In general, it's better for all involved to pick up after your dog.

What You Can Do

Many people think that if they make the stool taste bad, their dog will not eat it. There are powders you can purchase from the vet that are added to the dog's food and supposedly render the feces absolutely distasteful. I have found them to be ineffective. There are also home remedies such as sprinkling Tabasco sauce, cayenne pepper, or food additives directly onto the stool. These may actually become acquired tastes, and the dog will look forward to his next seasoned, fecal snack. Besides, it seems like a lot of trouble. Wouldn't it just be easier to pick up the poop? To stop coprophagia:

- *Don't give the dog the chance to eat feces.*
  Pick up every time he eliminates, and keep the cat box totally inaccessible to the dog.

- *Tell him what to do.*
  When you catch him in the act, tell him "off."

- *Distract the dog.*
  As soon as he is done eliminating, play ball with him, call him to you, and get really goofy with him. Do anything to keep him away from the stool.

- *Keep him stimulated.*
  If your dog is well exercised and mentally stimulated, he will not be bored. You decrease the odds when you fulfill your dog's other needs.

Compulsive Behaviors

Canines have their own version of obsessive-compulsive disorder (OCD). Some of the symptoms are dog equivalents to human symptoms. Acral lick dermatitis is the repetitive licking of a limb, which is similar to a person obsessively washing his hands. The following are other behaviors by which the disorder manifests itself:

- shadow chasing
- flank sucking
- tail chasing—the dog spins in circles trying to grab his tail

- fly snapping—the dog snaps at imaginary flies
- oral fixation—the dog wants something in his mouth at all times
- self-mutilation

Sometimes a behavior starts with a reason like a flea allergy and continues from there. Other times it is chemical. Fly snapping, for example, is thought to be related to a form of epilepsy.

Compulsive disorders in dogs are sometimes familial or genetic, and some breeds are much more susceptible than others. Anxious, hyperactive breeds, like Bull Terriers, or breeds bred to work closely with humans, like Labradors or Shepherds, often suffer from compulsive disorders. Dogs that form strong bonds and then suffer from separation anxiety when left alone for long periods of time will often develop a compulsive behavior, as will a dog with a lot of emotional baggage.

Treatment

Consult your veterinarian to treat compulsive disorders. There are several methods of handling a compulsive disorder, depending on what type of disorder the dog displays and how it originated. A low-protein diet, an increase in exercise, a more interesting environment, really good chew toys, and neutering are all possible solutions. Try to reduce or eliminate whatever conflict the animal is experiencing and avoid reinforcement of the behavior. Sometimes coddling can be detrimental. Depending on the situation, the solution may be pharmacological. Some of the drugs used to treat dogs are the same as the drugs used to treat humans with OCD, such as Prozac. If you have any concern about a compulsive disorder, please take your dog to the veterinarian. He may refer you to an applied animal behaviorist or veterinary behaviorist to oversee a treatment protocol.

And Remember . . .

Dog behaviors that serve a purpose in the dog world may be unacceptable in the human world. Dogs are predators that we bring into our homes. They are not guinea pigs that eat grass. They are social carnivores. We must teach them what we deem to be suitable behavior. If we stifle their natural behaviors completely, they will express themselves in other ways or when we are not around. Make sure you satisfy your dog's needs and provide proper outlets for his drives. In general, a well-exercised, highly stimulated dog is satiated. I will reiterate: a tired dog is a good dog.

In Short

Much of what pet owners react to as problem behaviors is thought to be no more than business as usual to the dog. When attempting to change problem behavior, remember to look at the big picture.

Mental stimulation and exercise greatly contribute to successfully solving behavior issues.

Separation anxiety is a serious issue for both the dog and the owner. The dog is in distress when separated from his pack. In an attempt to relieve his anxiety the dog may engage in destruction, uncontrollable house soiling, over-vocalization, and other potentially dangerous behaviors.

The following treatments often prove effective:

- Establish rules and boundaries immediately.
- Possibly confine the dog to a crate or other small area with comfort items like his bed and toys. Frequently enter and exit the room until he begins to relax.
- Come and go from the house. Gradually lengthen the amount of time you are gone.
- Practice low-key departures and arrivals. Ignore the dog fifteen minutes before you leave and after you return.

Chewing is another behavior that, when left unchanneled, causes damage and owner angst. Dogs need to chew. It is the owner's responsibility to provide proper outlets for the behavior with appropriate chew toys.

Like many other problems, inappropriate chewing is best solved by prevention through crating when the owner is gone and active supervision when the owner is at home.

There are many reasons dogs bark. Once you establish why the dog is barking, you can choose an appropriate solution.

To combat barking:

- Make sure the dog's needs are met.
- Issue a quiet command followed by a sound deterrent or breath spray.
- Reward the dog when he is quiet.
- Consider using a bark collar such as citronella.

Dogs primarily jump up to get closer to human faces when greeting people. One simple solution is to lower yourself to the dog's level for a welcome.

Other ways to diffuse a jumping problem are:

- making low-key arrivals and departures
- ignoring the dog until he settles down
- teaching him to sit when greeting people
- teaching the off command
- body blocking, or moving toward the dog into *his* space
- spritzing breath spray into the dog's mouth in conjunction with the "off" command

Digging is a natural behavior. Whether your dog is digging to relieve boredom, exterminate vermin, or cool himself off in moist soil, you want it to stop!

To stop digging, try:

- prevention—don't give him the chance to do it
- exercise—relieves boredom and releases energy
- the off command—tells him what to do when caught in the act

If running away is your dog's problem, or pleasure, work on strengthening both your relationship and your fence. If your dog is still intact, neutering alone may be the solution.

Pica, the eating of inanimate, nonfood objects, or coprophagia, the eating of fecal matter, may happen because of dietary deficiencies or, once again, boredom. Getting to the object or fecal matter before the dog is the key. Teaching the off command is particularly handy with these problems.

Shadow chasing, flank sucking, tail chasing, fly snapping, and other self-mutilations to the degree that the dog is causing himself bodily harm all fall into the obsessive-compulsive disorder category. This serious disorder often demands the skills of a behavior professional as well as pharmacological intervention.

Epilogue: A Final Note

One of the main reasons people bring dogs into their homes is the desire to have a relationship with another creature. Not only does owning a pet fulfill our need to nurture, it also provides us with a feeling of love and warmth. When you come home from a hard day at work, there can be nothing more comforting than the sight of your dog greeting you at the front door. Every time we interact with our dogs, we rediscover a part of ourselves that has been lost—a connection with nature and life.

Developing a loving, lifelong relationship with your dog takes time. It begins before you even bring the dog into your home and lasts forever in memories. Every minute you spend training and caring for your dog helps build the foundation for your relationship. As with everything in life, you can't skip steps. There are no shortcuts. Even when you hire a trainer or animal behaviorist, you need to be involved if you expect to have a solid relationship with your dog. It is the participation and the struggle that reap the joyous rewards of a true love relationship between you and your dog.

Now that Lotte is beginning to age, I remember how hard it was when Pearl's life began to end. No matter how difficult and painful the thought of losing Lotte, our loving relationship makes it all worthwhile. When the time comes, I will search high and low to find another companion for Ruby and myself. I now know that it is possible to have more than one true love in life.

Appendix

Quiz: Are You a Good Dog or a Bad Dog?

1. How does your dog respond to new people?
   a. fearful/aggressive
   b. timid
   c. mildly interested
   d. confident
2. How does your dog respond to new dogs?
   a. fearful/aggressive
   b. timid
   c. mildly interested
   d. confident
3. How often do you exercise your dog?
   a. rarely
   b. monthly
   c. weekly
   d. daily

4. How is the dog exercised?
   a. in the backyard
   b. short walks
   c. long walks
   d. vigorous workouts
5. How many hours a day is the dog left alone?
   a. more than ten
   b. eight to ten
   c. four to eight
   d. two to four
6. Where does the dog sleep?
   a. outside
   b. in bed
   c. loose
   d. crated
7. How is the dog fed?
   a. food left out
   b. fed whenever
   c. fed on a schedule
   d. works for food
8. When do you give the dog attention?
   a. dog nudges you
   b. you give it freely
   c. dog works for it
9. How do you train your dog?
   a. with corrections
   b. with food
   c. with both
10. How does your dog respond to your commands?
   a. ignores you
   b. hesitates but obeys
   c. obeys quickly

11. How do you reprimand your dog?
    a. after the fact
    b. during the behavior
12. What type of reprimand do you use?
    a. physical reprimands
    b. verbal reprimands
13. How do you play with your dog?
    a. dog initiates play
    b. you initiate play
14. Is your dog spayed or neutered?
    a. no
    b. yes

Scoring

Questions 1–7:
a) one point; b) two points; c) three points; d) four points
Question 8:
a) one point; b) two points; c) four points
Questions 9–10:
a) one point; b) three points; c) four points
Questions 11–14:
a) one point; b) four points

Results

Your dog's age, breed, and temperament all play a role in your answers.

43–56 *Great owner, happy dog:* You have a good understanding of the human-canine relationship. You give your dog love as well as boundaries and teach him how to live within our world.

29–42 *Good owner, content dog:* You are doing a good job raising your dog, but if you find you have troubles with compliance, maybe you should follow the rules of rank a little more closely. Or your dog may simply need more exercise and mental stimulation. You are only a few steps away from a strong leader-follower relationship.

14–28 *Fair owner, poor dog:* Because you love your dog, take the time to learn more about him and what is essential to him. Work on finding the balance between love and discipline, and you will be a good leader. Don't feel bad. With a little insight, you will be out of the doghouse.

Index

Accidents, housebreaking, 155–57, 163
Adolescence, 30
Advanced submission, 49
Age
  to begin fear therapy, 88
  to begin socialization, 74–78
  to begin training, 6–7, 29, 99–100
  at breeding, 24
  to bring a puppy home, 29
Aggressive dogs, 110, 113. *See also* Fear-induced aggression
Aggressive posture, 48
AKC Gazette, 24
Alpha dogs, 16, 44, 59–60. *See also* Dominant dogs
American Cocker Spaniels, 161
American Kennel Club (AKC), 23
Angel (dog), 134–35
Animal-learning theory, 98
Anti-chew cream and spray, 120
Appearance of dog, 20–21
Aroused posture, 48
Association of Pet Dog Trainers, 68

Babies, introducing dogs to, 133–35
Backyards, 192
Bark collars, 111, 202–3
Barking, 201–4, 214
  on command ("speak"), 203–4
Bassenjis, 194
Bassett Hounds, 136

Baths, 137
Becky (dog), 68
Beds, 121
Behavior problems, 191–215
barking, 201–4, 214
chewing. *See* Chewing
compulsiveness, 211–13, 215
coprophagia, 209–11, 214–15
digging, 99, 205–7, 214
exercise and mental stimulation for, 192
jumping, 204–5, 214
pica, 209, 214–15
running away, 207–8, 214
separation anxiety. *See* Separation anxiety
Behavioral science, 98
Beta dogs, 44
Bichon Frise, 161
Bingo (dog), 134–35
Bite inhibition, 37–39, 48
checking for, 32
"ouch" technique and, 38, 40
play and, 54
teaching, 38–39
time-outs and, 38–39, 40
Biting, 47, 131. *See also* Bite inhibition; Chewing; Mouthing
Blue (dog), xiii–xiv
Body massage, 140
Body posture/language
of dogs, 48–49, 65, 66
of humans, 106, 168
Boiling point, 50
Bolt snaps, 121
Bones
raw cow, 119, 196
sterile, 119, 196, 199
Booby traps, 208
Borchevski (dog), 158
Breed rescue organizations, 23–24
Breeders
finding, 23
questions to ask, 24–25
refusal to answer questions, 26
Breeds, 11–12, 15–27
cross, 19–20, 26–27
defined, 17
mongrels, 20, 26–27
natural dogs, 20
rare, 19, 26
researching, 22–26
Broken spirit, 100
Brushes, 119
Brushing, 136, 138
Buckethead (dog), 210
Buckle collars, 118, 123, 184
Buddy (dog), 68
Buddy Love (dog), 36
Bull Terriers, 54, 212

Bulldogs, 21, 136
Buster cubes, 119

Caan, James, 46
Canine Eye Registry Foundation (CERF), 24
Canine Good Citizen (CGC), 25
Car rides, 83, 86
Car sickness, 83
Challenging behavior, 65
Chew toys, 56
- appropriate, 196–98
- in crates, 126, 127
- inappropriate, 195–95
- keeping them interesting, 198–99
- types of, 119

Chewing, 8, 57, 195–200, 213. *See also* Biting; Mouthing
- prevention of inappropriate, 199–200
- reasons for, 55–56

Children, 74, 131–35, 141
Chloe (dog), 75–76, 88
Chocolate Cocker Spaniel, 65
Choke chains, 4
- alternatives to, 121–23

Citronella bark collars, 111, 202
Citronella barrier collars, 209
Citronella boundary collars, 111, 206–7
Citronella remote collars, 111
Clarke, Ross D., 23
Collars
- bark, 111, 202–3
- buckle, 118, 123, 184
- choke chain. *See* Choke chains
- citronella. *See* entries beginning with Citronella
- head halters, 111, 118, 122, 184
- leather vs. nylon, 123
- no-pull harnesses, 118, 122, 184
- nylon slip, 118
- pinch, 111, 118, 122–23, 184

Collies, 17
"Come" command, 168, 179–81, 185
Commands
- "come," 168, 179–81, 185
- confrontational, 183
- distance, 186–87
- double, 169–70, 187
- "down." *See* "Down" command
- "easy," 183, 184, 185
- enforcing, 65
- "go to your place," 187–88
- "no," 10, 104–5, 112, 205
- "off." *See* "Off" command
- potty. *See* Potty on command
- "sit." *See* "Sit" command
- "speak," 203–4
- "stay," 175–79
- "take it," 183
- during walks, 183–85

Communication, 53–54, 166–70

Compulsive behaviors, 211–13, 215
Conditioned reinforcement, 107, 114
Confrontational commands, 183
Coprophagia, 209–11, 214–15
Correction, 102–5, 108–13. *See also* Punishment after the fact; Reprimands
  appropriate, 109–12
  inappropriate, 109
  leash, 111, 170
  levels of, 110–11
  physical, 110–11
  power of, 112–13
Countercommanding, 87–88
Counterconditioning, 85–86, 90
Cradle position, 111, 137–38
Crates, 121, 124–31, 141
  alternatives to, 153
  appropriate situations for, 129–30
  bite inhibition and, 38–39
  in the car, 83, 86
  equipping, 127
  getting dog into, 127–28
  housebreaking and, 125, 151–53, 157, 162
  myths about, 9–10
  reasons for using, 125–26
  releasing dog from, 128
  reprimands in, 128–29
  size of, 126–27
  training dog to be alone in, 130–31
  types of, 126
Crossbreeds, 19–20, 26–27
Cypher, Julie, 134–35

Dachshunds, 89
Dalai Lama, 51
Democracy, 68
Dern, Laura, 36
Desire to please, lack of, 9
Diarrhea, 157
Diet. *See* Food
Digging, 99, 205–7, 214
Digging pits, 207
Dirty puppies, 22, 127, 152
Displays, 48–49
Distance commands, 186–87
Dog doors, 159
Dog license tags, 120
Dog parks, 78
Dominance levels, 48
Dominant dogs, 67–68. *See also* Alpha dogs
Dominant posture, 48
Double-commanding, 169–70, 187
"Down" command
  before feeding, 62
  inappropriate use of, 204, 205
  teaching, 174–75

Early fear, 49
Early submission, 49
Ears, touching, 139–40
"Easy" command, 183, 184, 185
Eating rights, 62
Ed (dog), 194
Electric mats, 208–9
Electronic bark collars, 111, 202–3
Elevation tests, 33
Entry protocol, 63
Etheridge, Melissa, 134–35
Euthanasia, xvi, 8
Evolution
  of dogs, 16–17
  of training, 97–99
Excitement urination, 160–61, 163
Exercise, 135–36, 159–60, 192
Exercise pens, 120
Expectations, unrealistic, 106
Extreme submission, 49

Family
  as pack members, 44
  rank of, 60
Fear, 72–74, 92
  coping with, 85–88
  levels of, 49
Fear-induced aggression, 73–74
  posture of, 49
Fences, 206–7
Fetch, 64
Flank sucking, 211
Flea combs, 119
Flip chips, 119, 196, 197
Flock guard dogs, 19, 26
Flooding, 85, 90
Fly snapping, 212
Fonda, Bridget, 186, 210
Food, 118. *See also* Treats
  accessories and types of, 118
  proper diet, 135
  scheduling of meals, 143, 147–49
  working for, 62, 106–8
Food bowls, 118, 132–33
Foot massage, 138–39
Foster, Jodie, 71, 89
Free-feeding, 62, 147–48

Games, 63, 64, 132. *See also* Play
Generalizing. *See* Situational learning
Genetic problems, 24
Gentling, 136–40
German Shepherds, 186, 210
"Go to your place" command, 187–88
Golden Retrievers, 19–20, 76, 149
Guilt, lack of, 11, 103
Gun (sporting) dogs, 19, 26

Handling, 136–40
Head halters, 111, 118, 122, 138, 184
Health problems, 156–57

Heat exhaustion, 136
Herding dogs, 19, 26, 54
Hierarchy. *See* Rank
Hip dysplasia, 24, 135
History, 5–6
Hounds, 19, 26
Housebreaking, 105, 143–63
  accidents and, 155–57, 163
  on command. *See* Potty on command
  confinement and, 149–53
  crates and, 125, 151–53, 157, 162
  dog doors and, 159
  dog signals in, 155
  habit in, 144–47
  paper training. *See* Paper training
  potty times and, 153–54, 162
  preferred locations in. *See* Potty areas
  preferred substances and, 145
  preparation for, 144
  puppy stations in, 150–51
  rain and, 146–47
  of reverse-housebroken dogs, 157–59
  scheduling of food and water, 143, 147–49
  submissive or excitement urination and, 160–61, 163
  umbilical cording and, 150
  walks for, 159–60
Huggie (dog), 21–22
Hunting instinct, 47
Hygiene items, 119

ID tags, 120
Ignoring bad behavior, 101, 110
Immunizations, 7–8, 76–78, 92–93
Independent puppies, 30, 40
Indulgence, 10
Insecure puppies, 30–31, 40
Instructive reprimands, 104–5, 112, 114, 171–72
Interested puppies, 30, 31, 40

Jack Russell Terriers, 19, 90
Jackpots, 108, 115
Jordy (dog), 68
Jumping, 204–5, 214

Kennelosis, 25, 89
Killer instinct, 46–47
Kong toys, 119, 196, 199
Krantz, Tony, 91

Labrador Retrievers, 91, 158, 196, 198, 212
Ladd, Diane, 36
Leader-follower relationship, 4–5, 5, 10, 45, 59–69. *See also* Rank
Leashes
  appropriate, 121
  corrections with, 111, 170
  holding correctly, 168–69

housebreaking and, 146, 149, 150
leather vs. nylon, 123
retracting, 121
sizes, 118–19
snap types, 121
Lemmon, Jack, 75–76, 88
"Let's go" command, 183–85
Linebreeding, 20–21, 24
Littermates, 91–92
Lotte (dog), xvi–xix, 15, 23, 33, 37–38, 53–54, 63, 68, 83, 107, 124–25, 145, 167, 203–4, 218
Love, praise, and affection fallacy, 4, 100
Lowell, Michael, 19
Lure-reward training, 107–8, 114

Man's Best Friend (film), 17
Marking, 51–52, 160
Maturity, 30
Medical and Genetic Aspects of Purebred Dogs (Clarke and Stainer), 23
Mental stimulation, 192
Michael, George, 22
Miles (dog), 91
Military-style training, 97–98, 114
Moderate submission, 49
Mongrels, 20, 26–27
Motion-sensitive noise devices, 208
Motivational training, 100–1, 114
Mousetraps, modified, 208
Mouth, touching, 139
Mouthing, 40, 55. *See also* Biting; Chewing
Movement, attraction to, 47–48
Muzzle grab, 33–36, 40, 109, 111, 129, 138
for barking, 201–2
for bite inhibition, 38
common errors when performing, 35–36
for jumping, 205
not for children, 132
performing, 34
Myers, Mike, 158
Myths, 6–11, 12–13

Nail clippers, 119
Name, teaching, 171–72
Natural dogs, 20
Negative pairing, 112, 205
Nell (film), 71–72, 85–86, 89, 91
Nell syndrome, 89
Neutering, 8–9, 46
Neuticles, 46
"No" command, 10, 104–5, 112, 205
Noise sensitivity, 32, 47–48, 86–87, 110
No-pull harnesses, 118, 122, 184
Nurturing, in first weeks of life, 25
Nylon slip collars, 118

Obsessive-compulsive disorder (OCD), 211–13, 215
Odor neutralizers, 120, 156
"Off" command, 168, 171, 200, 205
  teaching, 181–83
Omega dogs, 45
Oprah Winfrey Show, The, 65–66
Oral fixation, 212
Orthopedic Foundation Association (OFA), 24
Oscar (dog), 91
"Ouch" technique, 38, 40
Outcrossing, 24
Overbonding, 90–92, 193
Ovitz, Michael, 149

Pack behavior, 43–46
  family-oriented, 44
  in a multi-dog household, 65–68
Paper training, 6, 158–59, 163
Pass the Puppy (party game), 79–80
Paying attention, 169, 198
Pearl (dog), xiv–xvi, 15, 54, 218
Pencil test, 104
People, dogs compared with, 9
Personal puppy test, 30–33, 40
Pet-store dogs, 22
Physical manipulation, 111
Pica, 209, 214–15
Pig ears, 197
Pinch collars, 111, 118, 122–23, 184
Pit Bulls, 54, 198
Play, 53–55. *See also* Games
Play bow, 54–55
Potty areas, 82, 106, 144, 145–46
Potty boxes, 158
Potty on command, 82, 145, 154–55, 160, 163
Praise
  for chewing on toys, 199
  earning, 62–63
  housebreaking and, 146
  specific, 170
Predatory behavior, 46–51, 56–57
Prednisone, 157
Procreation, 45–46. *See also* Neutering; Spaying
Protein, 135
Puberty, 30
Pugs, 136
Punishment after the fact, 103–4, 109, 114, 156
Puppies, 29–40
  development of, 30
  dirty, 22, 127, 152
  holding, 137–38
  parenting, 39
  proper diet for, 135
  test for, 30–33, 40
  when to get, 29
Puppy classes, 75, 77, 105
Puppy gates, 121
Puppy parties, 78, 79–81
Puppy stations, 150–51

Quiz: Are You a Good Dog or a Bad Dog?, 219–22

Rain, 146–47
Rank, 16, 44–45, 59–65
  family in, 60
  in a multi-dog household, 66–68
  nothing for free in, 62, 64–65
  rules of, 60, 61–63, 69
  supporting hierarchy in, 67–68
Rare breeds, 19, 26
Raw cow bones, 119, 196
Rawhide, pressed, 119, 197–98
  softening, 199
Relationship. *See* Leader-follower relationship
Reno (dog), 46
Reprimands, 98–99. *See also* Correction
  for barking, 201
  from children, 132
  in the crate, 128–29
  instructive, 104–5, 112, 114, 171–72
REQUEST + ACTION = CONSEQUENCE, 102, 114
Rescue dogs, 11–12, 27, 193
  pros and cons of adopting, 21–22
  socialization of, 74–75
Rescue organizations, 23–24
Restraint tests, 33
Retracting leads, 121
Retrievers, 17
Reverse-housebroken dogs, 157–59
Revoking privileges, 66
Right Dog for You, The (Tortora), 19
Rottweilers, 19–20
Round-Robin Recall (party game), 80–81
Ruby (dog), 17–18, 30, 31, 51, 53, 54, 61, 63, 68, 78, 84, 110, 130, 167, 218
Running away, 207–8, 214

Sammy (dog), 161–62
Scent, 52–53, 195
Self-mutilation, 212
Separation anxiety, 11–12, 192–95, 212, 213
  barking and, 201
  preventing, 194–95
Separation reflex, 126, 130
Shadow (dog), 18
Shadow chasing, 211
Shampoo, 119
Sheedy, Ally, 68
Shepherds, 212
Sherpa bags, 78, 121
Sinatra, Frank and Barbara, 18
"Sit" command, 99, 103–4
  emergency distance, 187
  before feeding, 62
  teaching, 172–74
Situational learning, 105–6, 146

Sleeping rights, 61
Smooth Collies, xv–xvi
Sniffing behavior, 52–53
Socialization, 71–93
  by breeder, 25
  car rides and, 83, 86
  crashing parties and, 82
  defined, 72
  delayed, 75–76
  immunization vs., 76–78, 92–93
  importance of, 72–74
  littermates and, 90–92
  nature vs. nurture in, 88
  as ongoing process, 76
  problems and misconceptions, 88–92
  puppy parties and, 79–81
  of puppy vs. older dog, 89–90
  with resident vs. strange dogs, 90
  walks and, 81–82
  when to begin, 74–78
Solomon (dog), 65–66
Sophie (dog), 196
Sound. *See* Noise sensitivity
Spaying, 8–9
"Speak" command, 203–4
Sporting (gun) dogs, 19, 26
Spray Barrier, 209
Sprays, 111, 120, 128–29, 201
Spring snaps, 121
Stainer, Joan R., 23
Standard Poodles, 75
"Stay" command, 175–79
Sterile bones, 119, 196, 199
Strangers, introducing dog to, 81–82, 84
Street protection, 185
Streisand, Barbra, 161–62
Styptic sticks, 119
Submission levels, 49
Submissive posture, 48
Submissive urination, 160–61, 163
Subordinate dogs, 67
Sunflower (dog), 186
Surprises, preparing dog for, 140
Survival basics, 98–99
Systematic desensitization, 85–86, 90, 133, 137

Tail chasing, 211
"Take it" command, 183
Target test, 32
Teething, 8, 55
Television, 87
Temperament, 110, 112
Terriers, 19, 26, 54, 205, 207
Territory, 51–52
Thorne-Smith, Courtney, 194
Tibetan Mastiffs, 17, 51
Time-outs, 38–39, 40, 110
Timid dogs, 110, 113
Titles, 25
Tortora, Daniel F., 19
Toys, 63, 64, 196. *See also* Chew toys

Training paradox, 113–14
Training theory, 97–115
  author's school of, 101–13
  new school of, 100–1
  old school of, 99–100
Treats, 4, 12, 101–2, 106–8
  conditioned reinforcement with, 107, 114
  in fear therapy, 86
  handling and, 137
  housebreaking and, 154–55
  the jackpot, 108, 115
  lure-reward training and, 107–8, 114
  motivational training and, 100–1
  muzzle grab and, 35
  Pass the Puppy and, 79–80
  Round-Robin Recall and, 80–81
  socialization and, 74, 84
  "stay" command and, 176–77
  stuffing chew toys with, 199
  timing of, 108
  types of, 118
Tug-of-war, 64

Umbilical cording, 150
Urinary-tract infections, 157

Virtual Fence, 206–7

Walks
  commands during, 183–85
  elimination vs. exercise, 159–60
  socialization and, 81–82
Walter (dog), 21
Water, 143, 147–49
Weimaraners, 18
Winfrey, Oprah, 65–66
Withholding of rewards, 110

Your Purebred Puppy: A Buyer's Guide (Lowell), 19